Hollywoo

from the Silent Era to Film Noir

Film and Culture / John Belton, General Editor

Film and Culture / Edited by John Belton

A SERIES OF COLUMBIA UNIVERSITY PRESS

Patrick Keating

Hollywood Lighting
from the Silent Era to Film Noir

Columbia University Press   *New York*

Columbia University Press
Publishers Since 1893
New York   Chichester, West Sussex
Copyright © 2010 Columbia University Press

The author and Columbia University Press gratefully acknowledge the support of two grants
from the Trinity University Department of Academic Affairs and the
University Department of Communication in the publication of this book.

Library of Congress Cataloging-in-Publication Data

Keating, Patrick, 1970-
    Hollywood lighting from the silent era to film noir / Patrick Keating.
            p.   c.m
    Includes bibliographical references and index.
    ISBN 978-0-231-14902-0 (cloth : alk. paper)–ISBN 978-0-231-14903-7 (pbk. : alk. paper)–
ISBN 978-0-231-52020-1 (ebook)
    1.   Cinematography–Lighting.   2.   Hollywood (Los Angeles, Calif.)–History   1.   Title.

2009015406

Designed by Lisa Hamm

For my mother

Contents

Acknowledgments

When I was young, my parents, Maria and Dennis, bought me a VHS tape of *The Third Man*. Little did they know that I would watch the film repeatedly, subjecting the entire family, including my sisters, Coleen and Amy, to hours and hours of zither music. It wasn't the zither music that fascinated me—it was the cinematography. Without my family's patience and support, I never would have learned to love the art of lighting, and this book would not exist.

When I went to college, I already knew that I wanted to study film. Once at Yale, taking challenging, thought-provoking classes with Scott Bukatman, David Rodowick, and Angela Dalle Vacche, I began to think of film as a socially significant form of cultural expression. After graduating, I returned to my native Los Angeles and earned an M.F.A. in film production from the University of Southern California. USC gave me the opportunity to get behind the camera and study cinematography directly, under the guidance of cinematographers Woody Omens and Judy Irola. My thinking about film was further shaped by Ivan Passer, Bruce Block, and Rick Jewell. I also collaborated with several talented peers, including Jordan Hoffman, Dionne Bennett, Camille Landau, Michael Friedrich, Wei-shan Noel Yang, Benjamin Friedman, Mark Skoner, Chris Komives, and (outside USC) Daria Martin.

This book is based on the dissertation I wrote while at the University of Wisconsin–Madison. I had the ideal advisor for the project, David Bordwell, whose thorough knowledge of Hollywood film history is matched by his infectious enthusiasm for the art of film. David's work is a constant reminder that film studies can be both intellectually ambitious and enjoyable. I feel special gratitude toward Lea Jacobs and Vance Kepley, who nurtured this project in its early stages and served on my committee. Ben Singer shaped my thoughts about lighting in theater, painting, and photography. In addition, I was fortunate to be surrounded by great minds and great teachers, including Noël Carroll, Malcolm Turvey, Kristin Thompson, Ben Brewster, J. J. Murphy, Kelley Conway, Lester Hunt, and the late John Szarkowski.

In my teaching career I have received warm support from several colleagues, and I want to thank in particular Jeff Smith, Scott Bukatman, Kristi McKim, Bill Christ, and Jarrod Atchison. I also consider it a privilege to have taught hundreds of smart, engaging students, who have taught me much about the cinema.

I have discussed my ideas with many patient friends over the years. In addition to those already mentioned, I am grateful for the support of Vince Bohlinger, Jinhee Choi, Emily Cruse, Ben de Rubertis, Meraj Dhir, Lisa Dombrowski, Haden Guest, Warren Kim, Dara Matseoane-Peterssen, Mike Newman, Phil Sewell, Holly Willis, and Nancy Won. My family always reminds me that there is more to life than work—thanks, Kathleen, Mark, Aaron, Peter, Pat, Kristin, Hannah, Moira, Soren, and Isabella.

It has been a joy to work with everyone at Columbia University Press, especially John Belton, Jennifer Crewe, and Kerri Sullivan. James Naremore and the anonymous reader provided careful and encouraging responses to my manuscript. They caught several errors, and made useful suggestions for improvement. Scott Higgins thoughtfully answered questions about studying Technicolor, and Tom Kemper helped me to improve the writing of several chapters.

This project was supported by various grant awards from UW-Madison. An earlier version of the second chapter of this book, "From the Portrait to the Close-Up," received the Society of Cinema and Media Studies student writing award and appeared in *Cinema Journal*. I developed related thoughts about Hollywood cinema in articles published in *Aura* and *The Velvet Light Trap*. At Trinity University, I received two grants to help me cover the printing cost of the many illustrations included herein—one from the Department of Academic Affairs, and another from the Department of Communication. Thank you.

While living in Los Angeles and Madison, I took advantage of the local film culture, watching classic films in 35mm whenever possible. I saw many of the films discussed in this book on the big screen at UCLA and the UW Cinematheque, which offer superb film series every year. I conducted archival research at several institutions, watching Mary Pickford films and other silents at the George Eastman House; viewing several beautiful Technicolor films at the Academy Film Archive; studying both silent and color films at the Motion Picture, Broadcasting and Recorded Sound Division of the Library of Congress; examining rare films at the UCLA Film and Television Archive; and watching Warner Bros. classics at the Wisconsin Center for Film and Television Research. In addition to film viewing, my research has drawn on documents housed in the USC–Warner Bros. Archive, the USC Cinematic Arts Library, the UCLA Performing Arts Special Collections, the Academy of Motion Picture Arts and Sciences' Margaret Herrick Library, the Getty Research Institute, the New York Public Library, and the Harry Ransom Humanities Research Center at the University of Texas at Austin.

Finally, my warmest thanks are given to Lisa Jasinski. Lisa supported this project in so many ways—with her sense of humor, her emotional encouragement, and her sharp editorial eye. She made this a better book, and she makes me a better person.

Hollywood Lighting
from the Silent Era to Film Noir

Introduction: The Rhetoric of Light

> My opinion of a well-photographed film is one where you look at it, and come
> out, and forget that you've looked at a moving picture. You forget that you've
> seen any photography. Then you've succeeded. If they all come out talking
> about "Oh, that beautiful scenic thing here," I think you've killed the picture.
>
> —Arthur C. Miller, in *The Art of the Cinematographer*

Perhaps Arthur Miller did his job too well. Film is an art of light, but
the art of Hollywood lighting remains so subtle that it usually es-
capes our attention. This is unfortunate, because Miller and his
peers did much to shape our experience of the classical Hollywood cinema. Whether it is noticed or not, light can sharpen our attention, shift
our expectations, and shape our emotions. In this study I propose to look
closely at this art that was designed to go unseen and unnoticed.

Three questions will guide this study. First, what were the major lighting conventions in the Hollywood of the studio era? All film scholars are
familiar with Hollywood's three-point lighting system, but this was only
one of the countless techniques that constituted the art of Hollywood
lighting. A richly detailed account of the various approaches to lighting
should add greatly to our understanding of Hollywood cinema. Second,
what functions did those different conventions perform? Even the most
unobtrusive image can accomplish a range of tasks, from storytelling to
glamour, from expressivity to realism. Third, how did the discourse of
lighting shape these multifunctional conventions? The discourse of lighting is more than just a group of statements about arc lights and eye-
lights—it is also a set of *limitations* on what can be said about technology
and technique. These limitations made the use of certain lighting strategies seem inevitable, and others inconceivable.

Convention, function, discourse—this study will explain the ways
these three areas of cinematographic rhetoric are connected. We can
think of a lighting *convention* as a recommendation about what to do
given a certain cinematic context. When photographing a jail scene, consider shining a light through the bars to create a shadow. When photographing a woman, consider using a more frontal placement for the key-

light. When photographing a daytime interior, consider shining a light through the windows. Even the most pedestrian studio cinematographer knew hundreds of such conventions, as they offered him a basic starting point for arranging the lighting of any given scene.[1]

Some of these conventions were merely suggestions, one option among several. For instance: When lighting a murder scene, consider casting the murderer's shadow on the wall. No cinematographer was required to use this trick; however, the device appears in many different films, from many different studios, so it seems fair to say that this was a shared strategy that most cinematographers could choose to employ or ignore. By contrast, some conventions are mandatory. Example: Always use enough light to get an exposure. A cinematographer who chose to violate this rule was putting his job at risk. In between the optional guideline and the mandatory law we have a variety of default conventions. For instance: When lighting a person, use three-point lighting as a starting point. Cinematographers were not required to use three-point lighting in every scene, but it was a safe option in most circumstances.

We should resist the temptation to assign a fixed meaning to any one lighting technique. The same device can appear in dozens of different conventions, as in the following options: When you are photographing a night exterior, consider low-key lighting. When you are photographing a male character actor, consider low-key lighting. When you are photographing a crime scene, consider low-key lighting. The fact that low-key lighting can serve in multiple conventions means that its own significance is not fixed. Low-key lighting does not mean "sinister" lighting, and backlighting does not mean "angelic" lighting. Just as we need context to understand the meaning of a word, we need context to understand the meaning of a lighting technique.[2] Because a convention is a recommendation about what to do given a certain context, the context acts as a structuring component of the rule. Examining conventions, instead of individual techniques, will allow us to develop more precise functional explanations. This in turn will allow us to link conventions to the institutions that endorsed them.

The word "convention" can imply an arbitrary relationship between a technique and its function. However, we should remember that different techniques have different functional capabilities. If you want to emphasize the nuances of an actor's facial performance, a brightly lit close-up is simply a better tool than a distant silhouette, given that the former will allow the audience to see the actor's face, while the latter will not. Similarly, if you want to model the distinctive shape of a character-actor's face, a cross-light is simply a better tool than a frontal light, given that the former will produce more pronounced modeling shadows than the latter. The range of lighting conventions in which a technique is used must match its functional capabilities.

Still, Hollywood's lighting conventions were normative, in the sense that they encouraged greater standardization. Learning the trade involved learning the list of situations in which a particular set of devices could be used. In theory, we can imagine a cinematographer who never thought about why he was using any of his devices, but who nevertheless became an expert practitioner of the Hollywood style because he knew what to do when filming a jail scene, when filming a horror movie, and when filming a female star. In practice, most Hollywood cinematographers were certainly much more thoughtful than these examples suggest, but we must expect there to have been a certain amount of unthinking rule-following. Between getting all the scrims on all the lamps, all the flags on all the stands, and all the rolls in all the magazines, the cinematographer and his crew had too many practical obligations to allow them time for endless aesthetic contemplation. The functions were built into the rules; a cinematographer could accomplish a wide range of aesthetic functions merely by following the most popular conventions.

Evidence from trade publications like *American Cinematographer* suggests that cinematographers worked with four general groups of conventions: figure-lighting, effects-lighting, genre/scene, and composition. These are not mutually exclusive categories, but they provide a good starting point for understanding the practice of Hollywood lighting. The figure-lighting conventions consist of recommendations about how to light the performers. Three-point lighting was a valuable default option, but Hollywood cinematographers consistently varied this basic scheme in order to light men and women in systematically different ways. Cinematographers could also adapt their lighting schemes to accommodate long noses, short necks, or other "unusual" features. Effect-lighting was defined as "any type of lighting which attempts to reproduce the effect of the illumination you'd actually see in any particular room or place under the conditions of the story."[3] In other words, a skillful cinematographer knew how to duplicate the look of sunlight shining through a set of blinds, how to create the effect of moonlight peeking through the trees, and how to set up a shot that appeared to be dimly lit by a single light bulb. The genre/scene conventions consist of recommendations on how to light particular kinds of story situations. Most of the rules are simple enough: comedies are bright; dramas are somber; romances are soft; crime stories are hard. However, applying these simple rules could be surprisingly difficult, given that most Hollywood stories combine elements of different genres. An expert Hollywood cinematographer had to know how to make a smooth transition from a comic love scene to a brutal crime scene. And, finally, a composition convention is a recommendation on how to organize the lighting of the complete picture. The most basic composition convention was the rule about putting the brightest light on the most important part of the story, but composition conventions were not entirely driven by storytelling considerations. A well-composed picture would also have aesthetic virtues like dynamism and balance.

If it seems as if Hollywood cinematographers had a lot of rules to follow, this is because they did. In fact, Hollywood cinematographers had so many conventions to draw from that they could not possibly follow all of them at the same time. For instance, what do you do when you need to photograph a woman in a crime scene? The figure-lighting conventions would call for soft lighting, but the genre/scene conventions would call for hard lighting. This problem—call it the problem of conflicting conventions—helps account for the complexity of Hollywood lighting techniques and conventions. Each Hollywood film posed a unique problem to the cinematographer, because each Hollywood film called for the use of a slightly different combination of conventions. The art of Hollywood lighting was an art of balance, requiring cinematographers to weigh the needs of the star against the needs of the story, or the needs of the set against the needs of the shot.

None of these conventions was created out of an interest in needless complexity; they were all created to help cinematographers achieve certain functions. Here *American Cinematographer* is an invaluable source. The discourse of this journal demonstrates both the normative and the functional quality of Hollywood lighting. On the one hand, the journal standardizes the Hollywood style by disseminating the various guidelines. Its discourse helps create norms. On the other hand, the journal does not enforce convention by fiat. Rather, it encourages the widespread adoption of a particular convention by reminding filmmakers of its functional benefits. Most cinematographers did not use backlighting because of unthinking obedience to an ironclad rule; they used backlighting because they believed that it could separate an actor from the background. While blind rule-following was a theoretical possibility and occasionally a practical necessity, the conventions were ultimately sustained because filmmakers continued to believe that certain devices performed certain functions, and that those functions had value.

As with the conventions, we can organize the *functions* of lighting into a few general categories. The first category could include storytelling functions, such as picking the protagonist out of a crowd, indicating the time of the scene, or setting the mood for romance. Because storytelling includes tasks like directing the audience's attention, the Hollywood lighting style is sometimes considered a "neutral" style, with "expressive" lighting limited to certain minor genres like the crime film and the horror picture. One of my goals in this book is to show that the "neutrality" account is mistaken. From the crime film to the romance, from the twenties to the forties, expressive lighting was a staple of Hollywood cinematography. Setting the mood was one of the best ways to tell the story.

The American Society of Cinematographers (ASC) also encouraged cinematographers to create "realism" with their lighting. Of course, "realism" is a famously amorphous concept, and cinematographers interpreted the term in complicated ways. Sometimes they meant that lighting should create an illusion of roundness, allowing spectators to see depth in a two-dimensional image. Sometimes they believed that cinematography could create a more powerful kind of illusion—the illusion of presence, whereby spectators come to experience the film as an ideal observer located within the fictional world. Realism could also refer to a particular mood, a harsh atmosphere that sets the tone for gangster films and social problem dramas. Perhaps the most important kind of realism was the realism of detail. Many cinematographers took pride in their ability to mimic the precise look of sunlight as it slants through a window, or the exact image of moonlight as it bounces off a lake. We should not assume that these four senses of the term "realism" reinforce each other; for instance, a conspicuously accurate lighting effect might draw attention to itself, thereby undermining the illusion of presence.

In addition to being storytellers, Hollywood cinematographers thought of themselves as image-makers. As such, they consistently aspired to certain ideals of pictorial quality, above and beyond the needs of the particular story. While many cinematographers believed that conspicuous beauty was distracting, even the most unobtrusively oriented cinematographers agreed that all images had to meet certain minimum standards of balance and proportion. Some cinematographers pushed the search for beauty even farther. The ASC encouraged such explorations by defining cinematography as "painting with light."

In contrast to the aesthetic agenda of the ASC, studio bosses like Irving Thalberg and Harry Cohn promoted a different ideal: glamour. While glamour could support the story by setting the right mood for romance, we might more usefully think of glamour as an independent imperative that may or may not overlap with the demands of storytelling. When a star is playing a handsome, romantic playboy, and he is lit like a handsome, romantic playboy, the lighting serves storytelling and glamour at the same time. When a star is playing a wretched, starving peasant, and he is still lit like a handsome, romantic playboy, the lighting serves glamour at the price of storytelling. The mandates can converge or clash.

These functional groupings—storytelling, realism, pictorial quality, and glamour—do not provide a comprehensive theory of lighting; rather, they indicate the purposes for lighting that seemed most relevant to Hollywood cinematographers. It is tempting to look for correspondences between the four sets of conventions and the four sets of functions. Should we say that the genre/scene conventions are geared toward storytelling? How about a link between figure-lighting conventions and glamour? Perhaps effects-lighting conventions are conventions of realism? Wouldn't composition conventions be conventions of pictorial quality? While these are intriguing correspondences, we should resist the temptation to overemphasize these links, for a very important reason: most Hollywood lighting conventions are multifunctional.

For instance, consider the genre/scene convention calling for crime stories to be shot in low-key lighting. We might point out that this convention sets the right mood for the story, and leave the analysis at that. However, we should also note that this strategy will provide appropriate character lighting for the tough-guy leads, add realistic detail to the underworld locations, and invoke an established pictorial tradition of representing crime. The convention fulfills several different functions at the same time. Indeed, it seems likely that the most popular conventions were widespread precisely because they were so multifunctional. Three-point lighting directs our attention to the actor's face (which aids storytelling), while making the actor look attractive (which enhances the star's glamour), while also creating an illusion of roundness (which is a kind of realism), and ensuring a pleasing balance of light and dark (which creates a level of pictorial quality). This helps to explain why three-point lighting became a default convention, rather than merely one option among many equals.

The concept of multifunctionalism has a great deal of explanatory power, allowing us to understand why certain conventions became so prevalent. Still, we should again remember that there were limits to the number of functions that any convention could perform. Some conventions were simply better than others at fulfilling the glamour function, and some conventions were simply better than others at fulfilling the storytelling function. This brings us back to the problem of conflicting conventions. As flexible as the lighting conventions were, they would often come into conflict with each other. In such cases, a cinematographer had to decide which convention he wanted to favor. It makes sense to suppose that a cinematographer would base the decision on functional grounds. If the situation called for more glamour, the cinematographer might favor a convention that performed the glamour function. If the situation called for greater narrational clarity, the cinematographer might favor a convention that performed a storytelling function. This would seem to solve the problem, but it actually makes it more complicated, because now we have introduced questions of value. When was glamour more valuable than storytelling? When was pictorial quality more valuable than realism?

Some scholars would argue that Hollywood filmmakers generally answered these questions by following the ideals of the "classical" style. For instance, David Bordwell, Janet Staiger, and Kristin Thompson argue that the classical Hollywood cinema is committed, above all, to one function in particular: the function of clear narration. Cinematographic style usually works unobtrusively, carefully directing the spectator's attention to the unfolding causal chain.[4] Other scholars offer different accounts of the classical style, placing added emphasis on the idea of the "invisible style" or stressing the importance of the illusion of presence.[5] Still other critics argue that Hollywood was not classical at all. For instance, Richard Maltby argues that Hollywood's multiple functions were always in conflict, with no single ideal assuming a dominant position. Storytelling is dominant in some scenes, but in many scenes other functions come to the fore.[6]

These debates concern the nature of Hollywood's aesthetic norms—whether they were classical or not, and whether that classicism was consistent. There is another important debate, about the best way to *explain* those norms—why filmmakers adopted classical (or nonclassical) norms, and why they changed when they did. Bordwell, Staiger, and Thompson explain the norms by pointing to the role of institutions like the Academy of Motion Picture Arts and Sciences, the ASC, and the Society of Motion Picture Engineers (SMPE), which provided forums for the discussion and dissemination of the Hollywood style, turning generalized aesthetic ideals into practical norms.[7] Mike Cormack takes a different approach, shifting the focus from institutions to ideology. In an analysis of Hollywood cinematography of the thirties, he argues that cinematographic style changed over the course of the decade, moving from an unpredictable, somber style that expressed a sense of crisis to a more restrained, high-key style that expressed the American ideology's reassertion of faith in the capitalist system.[8]

Another historian, James Lastra, has proposed a particularly rich methodology, combining an interest in institutions with a concern for ideology. Specifically, Lastra has studied the discourse of the Society of Motion Picture Engineers before and after the transition to sound.[9] While drawing on the work of Bordwell, Staiger, and Thompson, Lastra introduces a notable shift in emphasis. Hollywood was not a single institution; it was composed of several different institutions with interests that were simultaneously competing and overlapping. These institutions shared a general set of principles (such as the commitment to storytelling), but each institution may have interpreted those principles in different ways. In order to understand Hollywood as a whole, we need to pay more attention to these individual institutions, with their specific agendas, their specific cultures, and their specific ideals.

We might distinguish between these scholarly approaches by noting that their explanations can draw links among three different levels: 1) the level of film style, 2) the level of Hollywood institutions, and 3) the level of the broader cultures wherein the films were produced and consumed.[10] In *The Classical Hollywood Cinema*, Bordwell, Staiger, and Thompson are primarily interested in explaining the relationship between the first two levels, showing how shared norms have been shaped by Hollywood institutions. By contrast, Lastra is not very concerned with questions of film style; instead, he skillfully explains the cultural roots of various institutional conflicts, thereby drawing links between levels two and three. Meanwhile, Cormack hopes to explain the way ideology shaped the development of film conventions, but he is less interested in explaining the role of institutions as mediators of this relationship.

My own approach is firmly grounded in the tradition of "historical poetics," which seeks to explain the larger principles used in the construction of films. Like Bordwell, Staiger, and Thompson, my primary focus will be on the relationship between the conventions of Hollywood lighting and the institutions that produced those conventions. While taking advantage of this approach's strengths—such as its fine-grained attention to stylistic detail, and its commitment to precise historical explanations—I will also draw insights from other sources, such as Maltby and Lastra. *The Classical Hollywood Cinema*'s attention to the nuances of style is absent from Lastra's approach, but he compellingly demonstrates the benefits of attending to institutional and discursive conflict. The ASC had developed their own culturally specific discourses of light—discourses that would often conflict with the ideals of other Hollywood institutions. Because of this emphasis on conflict, my account of Hollywood lighting will sometimes echo Maltby in placing special emphasis on the ways that cinematographers offered an idiosyncratic and occasionally unclassical interpretation of the basic Hollywood norms.

With this approach in mind, how can we characterize the aesthetic ideals of Hollywood cinematographers? We should not assume that an endorsement of one classical principle implies an endorsement of them all. It is possible to endorse storytelling as an ideal without necessarily adopting the concept of the invisible style. Similarly, it is possible to embrace the concept of the invisible style without necessarily accepting the concept of the illusion of presence. The links between these ideals changed over time. During the silent period the ASC worked to construct a new identity for the cinematographer, not as a manual laborer but as a man whose artistic talents were just as impressive as his technical skills. In order to construct that identity, the ASC promoted an eclectic set of aesthetic ideals. Lighting could clarify the story, establish the mood, enhance characterization, add realistic detail, and astonish the eye with pictorial beauty. Most cinematographers were indeed committed to the ideal of storytelling, but they believed that storytelling was perfectly compatible with extravagant displays of spectacle.

After the transition to sound, the discourse of the ASC grew more classical, linking the ideals of storytelling, invisibility, and the illusion of presence. Cinematographers were much more likely to take pride in unobtrusiveness. However, this shift in thinking did not result in a more stable or standardized system. Within the ASC there were differences of opinion about the best ways to use style in support of the story. Meanwhile, cinematographers came into conflict with studio bosses, who demanded glamorous photography at all times. In short, the art of lighting was always the site of tension between classical and nonclassical principles, though the nature of that tension changed over the years.

While this approach places a great deal of emphasis on the notion of struggle, we must remember that most struggles do not end in a state of chaos. Instead, struggles lead to compromises, exchanges, and even some mutually beneficial solutions. It might be useful to imagine the Hollywood institutions as a set of loosely overlapping circles. The ASC, the SMPE, and all the other institutions of Hollywood shared some common interests—most obviously, the financial health of the industry as a whole. However, their interests did not completely coincide, and they would occasionally promote different kinds of ideals. Sometimes this would lead to unresolved tensions, but the conflict could just as easily lead to the discovery of some surprisingly elegant compromises.

In this analysis of Hollywood's rhetoric of light, I want to steer a middle course, between an emphasis on unanimity and an emphasis on fragmentation. Hollywood institutions had a shared set of interests, but this does not preclude the possibility of power struggles between institutions, and within institutions. In general, the institutional agendas almost overlap; the functional ideals almost coincide; the lighting conventions almost interlock. This study will attempt to define these three "almosts" with greater precision.

My analysis covers the period from 1917 to 1950. Bordwell, Staiger, and Thompson have proposed 1917 as a plausible date for the beginning of the classical Hollywood cinema. By this time Hollywood companies were committed to a certain product—the feature-length narrative film—and Hollywood filmmakers were committed to a certain set of conventions—the rules of the continuity system. Meanwhile, more and more cinematographers were adopting three-point lighting, which would soon become the signature lighting style of the Hollywood cinema. There is also an institutional reason for starting in the latter part of this decade: the ASC was chartered in 1919, and *American Cinematographer* began publication in 1920. The ending date is more open; one could argue that many of the conventions established in the studio period are still relevant today. Nevertheless, 1950 provides a suitable ending point, around the time Hollywood experienced a series of major changes, such as the "Paramount Case," the rise of television, the retirement of several veteran cinematographers, the transition to widescreen, and the adoption of new color film stock.

This book has three major parts, which cover, roughly, the twenties, the thirties, and the forties. The three parts will develop the argument in three distinct stages. Part I will explain the development of the rhetoric of light, arguing that the style of the silent period was not yet classical in the full sense of the term. No longer a laborer turning a crank, the cinematographer had become a skilled professional making a valuable artistic contribution to the cinema. At first, cinematographers defined the art of cinematography in different ways, drawing on various artforms, such as portrait photography and the theater. The result was an eclectic set of practices that incorporated some classical norms while ignoring others.

Part II will argue that the discourse and practice of Hollywood lighting became more classical after the transition to sound. Cinematographers were more likely to uphold unobtrusiveness as an important ideal, and the idea of the illusion of presence became a commonplace. On a practical level, lighting became highly conventionalized, with an assortment of conventions both allowing and encouraging the cinematographer to vary his style to suit a wide range of factors, from the story to the star, from the setting to the scene. Lighting may have been unobtrusive, but it was never neutral. All of these rules birthed a new problem: following one rule might require the cinematographer to break another. Cinematography became an art of balance, as cinematographers figured out creative ways to fulfill competing mandates, serving the story and glamorizing the star at the same time.

With the working principles of classical lighting established in part II, part III will consider different variations and departures from the classical norms. In the late thirties and the forties, cinematographers worked to incorporate new devices, such as Technicolor and deep-focus cinematography. Although some cinematographers succeeded in assimilating these techniques to the classical system, many took the opportunity to explore new creative possibilities. Indeed, we can draw a useful distinction between two different approaches: the classical, which sought to balance competing functional priorities, and the mannerist, which pushed the limits of the Hollywood style by intensifying the impact of a particular function—making the film more expressive, more realistic, or more pictorial.

In short, part I will look at the classical Hollywood style before it became fully classical; part II will examine Hollywood lighting in its most classical period; and part III will consider different variations and departures from the classical norms. Throughout these sections, my goal is to identify the conflicts and convergences that made the art of Hollywood lighting one of nuance and detail.

In conclusion, a brief anecdote might summarize these ambitions. The longtime Metro-Goldwyn-Mayer cinematographer Joseph Ruttenberg once told a story about a black-and-white cinematographer who liked to keep two small light bulbs underneath the camera: one red, one green. An actor once asked the cinematographer what the bulbs were for, and he received this reply: "If it's a dramatic scene, I put the green bulb there."[11] Ruttenberg does not offer any additional explanation, but we might speculate that the green bulb was slightly less bright than the red bulb, producing a slightly less brilliant eye-light, which was ever so slightly more appropriate to the mood of the dramatic scene.

Most cinematographers were capable of making spectacular images of astonishing beauty, but we should not forget that the art of lighting was usually in the details. This book will take a closer look at those details, keeping in mind that the art of lighting is all-pervasive, influential in even the smallest changes to the tiniest points of light.

Part I

Lighting in the Silent Period

Mechanics or Artists?

In 1917 *The Moving Picture World* published a special issue covering all the major fields of filmmaking. In the largest article on cinematography, William Fildew wrote, "As to what constitutes the greatest difficulty in the making of motion pictures, I should reply the insecurity of the tripod in the making of outdoor scenes. . . . The tripod must be nursed like a contrary child. It must be firmly set."[1] For Fildew, cinematography posed a mechanical problem, requiring a mechanical solution.

Ten years later, the same magazine ran a story about cinematographer Joseph La Shelle. According to the article: "During the past few years the motion picture cameraman has come into his own to a certain extent. Producers began to realize that a camera was not merely a contraption with a crank and a reel of film but to the expert cameraman represented the brush and easel of the portrait artist."[2] The metaphor of the parent nursing a child was replaced with a new conceptual definition: the cinematographer as artist. Soon this metaphor would become a cliché. Cinematography was painting with light.

The point of making this comparison is not to say that cinematography in 1927 was aesthetically superior to cinematography in 1917. There were several ambitious cinematographers working in the late teens, and the cinematographer of 1927 still had to prevent his camera from tumbling down a hillside. Rather, I wish to suggest that the cinematographer had acquired a new public identity. He had come to be perceived as a person with good taste, emotional sensitivity, and a deep understanding of dramatic values. This new understanding was promoted by a particular institution: the American Society of Cinematographers. Through its monthly journal, *American Cinematographer*, the ASC crafted a compelling narrative about the development of a new kind of art—and a new kind of artist. No longer a laborer turning a crank, the cinematographer was a skilled professional making a valuable contribution to the cinema—a contribution that could best be described as aesthetic.

In 1913 the professional cameramen of Los Angeles and New York City formed two different clubs—the Static Club in California, and the Cinema Camera Club in New York. The clubs sponsored dances and other social events, but their primary purpose was to promote the interests of cameramen. The New York club's constitution read:

> We, the members of the Cinema Camera Club, have resolved to organize an association for the development of an artistic and skillful profession, namely, the operating of cinematographic cameras; it being our purpose to maintain for the members of said profession the dignified standing justly merited, among the rest of the industry of which it forms a most important branch.[3]

The club's members aimed to improve the professional standing of the cameraman by emphasizing his identity as a skilled artist.

A few years later the Static Club changed its name to the Cinema Camera Club, as the two clubs grew more closely affiliated. In 1918 Philip Rosen, formerly a member of the New York club, took over the leadership of the Los Angeles club. Soon this club became the American Society of Cinematographers. The ASC was an advocate for cinematographers, but it was not a union. Membership was by invitation only, and ASC members were regarded as elite cameramen. Although Rosen himself became a director, many of the other original members—including Charles Rosher, Joseph August, and Victor Milner—went on to have long, distinguished careers in cinematography.

Perhaps the first example of ASC discourse was its motto: "Loyalty, Progress, Art." Implicit in these three words is a certain narrative: as more and more aesthetic functions (art) are shared (loyalty), the group style improves (progress). This narrative would be refined in the discourse of the ASC's most visible publication, *American Cinematographer*. The first issue of this magazine appeared on November 1, 1920. Its masthead described the journal as "an educational and instructive publication espousing progress and art in Motion Picture Photography, while fostering the industry."[4] As the terms "educational and instructive" suggest, one of the journal's purposes was to help standardize Hollywood style by teaching all cinematographers—both amateur and professional—a set of normative ideals.

The desire to appeal to a wide readership was stated explicitly in the July 1, 1922, issue of the magazine: "While this magazine has already gained recognition throughout the industry as the only publication of its kind, the board of directors has prepared a program of expansion that will make The American Cinematographer of even more general appeal and greater national influence."[5] The strategy is clear: by enhancing public awareness of the cinematographer, the ASC hoped to increase his influence.

Because of its explicit concern with public identity, we should not expect *American Cinematographer* to offer a perfectly accurate picture of the cinematographer. Instead, we should expect its portraits to be as flattering as possible, like the soft-focus glamour shots that the ASC's members made for Hollywood stars. Just as the glowing highlights of those glamour shots are occasionally lacking in plausibility, the glowing reviews found in *American Cinematographer* are often lacking in consistency. One page may stress efficiency as the most important virtue; the next page may stress technical skill; and the next page may stress artistry. Indeed, its editorial policy was highly eclectic. Most articles offer practical advice, such as a 1921 article called "Composition—What Is It?" Some articles advocate policy changes advantageous to the cinematographer, as in the sensible article calling to "Eliminate Death from Air Cinematography." Other articles read more like wish-fulfillment fantasies, such as the optimistic 1926 article "Amateur Cinematography as World Peace Agent." Finally, a few articles seem to exist solely as filler. One of the magazine's first editors was a theosophist, and his issues occasionally include mystical articles, such as the 1928 piece "Significance of Jewels: The Influence of Rock Crystals and Other Stones on Mankind Said to Be Anything But a Dream."[6] There is nothing surprising about this inconsistency. Institutional discourse is often opportunistic. We should expect the ASC to use any argument that could highlight its guiding theme: why the cinematographer deserves an improved standing in the industry.

The fact that the discourse aimed to flatter the cinematographer does not mean that it should be discounted. Quite the contrary—the fact that it offers an *idealized* presentation of cinematography makes such writing a useful source for an analysis of cinematographers' *ideals*. Consider, for example, the series of comic articles signed by a fictional character named "Jimmy the Assistant."[7] These essays are invaluable because they explicitly foreground issues of class identity. In so doing, they help us understand how the desire to enter the professional classes helped shape cinematographic practices. Like many cinematographers (particularly the ones who were formerly assistants), the fictional Jimmy comes from the working class, but (unlike the members of the ASC) he seems to be eternally trapped in this status. In one article, he writes:

> Even guys that used to be camera punks and drop magazines right along with me has had their spell of being a first and some of them is even setting in a director's chair with a deep voice and a megaphone to use it through; but as for me—I'm still picking 'em up and laying 'em down—Jimmy the Camera Punk—always was and never will be. I got as much chance of being anything else as Will Hayes [sic] and Welford Beaton has of kissing.[8]

Jimmy's articles are peppered with jokes, slang, and misspelled words— he even gets Motion Picture Producers and Distributors of America (MP-PDA) head Will Hays's name wrong. His insights, though, are sharp: Beaton, a noted critic of Hollywood, was surely one of Hays's least favorite people. Some of the articles include a picture of Jimmy, represented as an anonymous laborer struggling under the weight of the assistant's regular load. Although Jimmy's persona is calculated to appeal to cinematographers who worked their way up from being assistants, his real function is not to remind them of their humble roots, but to flatter them about how far they have come. The cinematographer may have been a hard-working assistant once, but he is now something more valuable.

How did the discourse of cinematography define that value? Jimmy offers one answer in an article entitled "Wages and Salaries":

> Here I am, an assistant; I lug the camera junk around and hold a slate and do all the hard work there is around a camera. It's a hard graft and I get wages.
>
> The cameraman I work for drifts into the studio two minutes before the hour on the call, gives orders all day, takes orders from nobody, does pretty much as he pleases, and he gets a salary. I do about 10 times as much labor as he does and he gets about 15 times as much dough as I do.
>
> Now, let's figure it out. I do the most work, but he's worth a lot more than fifty times as much as me to the company. Fifty of me couldn'ts [sic] get the results that one of him gets. He gets a big salary but he earns many times that in not having retakes, doing good work under bum conditions, and in time saved by fast, efficient work.[9]

Here Jimmy measures value in purely practical terms, as a matter of speed. It is his highly trained efficiency that elevates the cinematographer above Jimmy's working-class status and into the professional class.

Jimmy, the champion of efficiency, sometimes argues that art is a pretentious waste of time. In one article he writes:

> There was a first cameraman come to the lot, a forrener from Checkovia or somewhere, and he talked English like a excited Chinaman trying to recite a Bjornsjern-Bjornsjern poem backwards with the original language. He didn't make sense. . . .
>
> We got out and started, and you should have seen this guy do his art. He was slower than the next raise. I had to laugh when I see him figuring exposure. He had a actinometer with more different gadgets on it than a flute has toots, and everytime he'd work it he'd forget something and have to start all over. All this got over big with the boss. He was getting art at last. Then this guy sets his stops as careful and delicate as a blind man peeling a cactus. Finally we gets the scene, and a few more before dark.[10]

This is vicious satire, with a strong and unpleasant dose of xenophobia, as Jimmy appeals to the ugly anti-immigrant sentiment that was disturbingly commonplace in the twenties. Jimmy's American slang is supposed to capture the voice of common sense, while lofty artistry is associated with a Slavic-Chinese-Scandinavian jumble of foreignness.

Many articles in the journal sound the same theme (albeit without Jimmy's stinging wit): the cinematographer is not just a laborer—he is a professional with a special gift for efficiency. Still, this emphasis on quick, competent work was probably not enough to convince producers of the cinematographer's worth. During the silent period, the cinematographer physically turned the crank on the camera, and, to some producers, an efficient cinematographer was still just an efficient crank-turner. In the words of one cinematographer: "There are a great many producers who affirm that cinematography is not an art, and the photographer therefore not an artist, but a mechanic just like the electrician and carpenter, and therefore not entitled to the salary he receives."[11] To counter this prejudice, *American Cinematographer* had to turn to a different kind of discourse: the discourse of Art.

Even Jimmy supports the aestheticization of the cinema. In one article he suggests that directors take inspiration from classics in the other arts, such as the music of "this Frenchman, Debuzzy," and Dante's story of "Poolo and Francesca."[12] In another article, entitled "Art and Business," Jimmy explicitly argues that the values of art and efficiency can be combined. According to Jimmy, there are three kinds of cinematographers:

> There's some cameramen which gets along great because although mebbe there work aint up to the highest artistick standard or aint A number 1 in a lotta ways, yet they're quick as lightnin' and dont never hardly ever get into greef in spite of there speed. They can cut from three to five days off a perduction filming time. That boosts him way up in the money saving class. . . .
>
> Then there's the other kind which is so downright klever that they can do most anything with a camera when it comes to turning out fine work. These birds is almost always slow. They has to be. You can't turn out marvellous work all the time like you was grindin corn. It takes thought and originallity and both takes time. But these guys work is so wonderful when they do get it that it dont make no difference how long it takes, because it's worth it to the picture. The extra time is made up in extra fine work, and pays a bigger return to the compeny than poorer faster photography wood. All five of these guys is workin' on long term contrackts. . . .
>
> The cameramen most in demand is the kinds whitch is a good cross between the two classes mentchioned. And that just nacherally calls for a pretty good mixture of business and art.[13]

The efficient cameraman saves money; the artistic cameraman makes money. The cameraman who manages to combine efficiency and artistry offers the best value to the producer.

This sounds like an ideal solution, but Jimmy was hardly the ideal spokesman for this position. If producers did not believe that cinematographers were artists, they were not going to be convinced by a fictional character who spelled Debussy "Debuzzy." The case was more compelling when it came from a cultured Frenchman like Joseph A. Dubray, ASC, the magazine's technical editor. Educated in Milan, Dubray shot such films as the 1925 version of *The Awful Truth* before taking a job at Bell and Howell. In a 1922 article Dubray writes:

> The cinematographer is now-a-days a recognized creative entity; he is cultured, studious, has high moral and artistic ideals; and by his intelligence he has gratified the screen with many remarkable achievements.
>
> The work of the cinematographer is complex.

Not only good photography is required from him, but also an artistic personality, a good knowledge of dramatic construction, and of the psychological value of the graphic presentation of a dramatic or comical situation.

This work cannot be considered as "labor," but by its nature, the responsibilities inherent in it, the mental effort required to carry it on successfully, classes it among professions, and it should be remunerated accordingly.[14]

Dubray and Jimmy are working toward the same goal, but employ two different discursive strategies. Jimmy offers a convincing argument that the cinematographer does professional work, but Jimmy's working-class identity only serves to remind readers that most cinematographers never received official training in the arts. By contrast, Dubray argues that the cinematographer is a sophisticated intellectual—an elite, cultured man whose work should never be confused with "labor." In attempting to win the cinematographer a new class status, Dubray fashions for him a new class identity.

The ASC was hardly the first to suggest that cinematography could be artful. In 1910 several articles in *The Moving Picture World* had proposed a distinction between moving photographs and moving pictures, with the former denigrated as mechanical reproductions and the latter celebrated as artistic creations.[15] A few years later, the same journal published a series of articles by Thanhouser cinematographer Carl Louis Gregory, who complained that cameramen were too often classified as mechanics rather than artists.[16] A rival publication, *Wid's Films and Film Folk*, routinely praised films that featured beautiful imagery.[17] Meanwhile, directors like D. W. Griffith, Cecil B. DeMille, and Maurice Tourneur were all developing self-consciously artistic visual styles. In many of these cases, the rhetoric of artistry performs an ideological function, making films more appealing to middle-class audiences who were concerned about the industry's popularity among working-class, immigrant audiences. For instance, one of the early *Moving Picture World* articles invokes the ideals of "uplift" and "progress," suggesting that pictorialism will elevate the tastes of film audiences.[18] Another article worries that inartistic films will simply pander to the tastes of "a certain section of New York City"—presumably, the Lower East Side.[19] Historians do not agree when this campaign for the middle-class audience (and against the working-class audience) ended, but it was certainly over by the time of the twenties. For this reason, we can say that cinematographers were not necessarily interested in adopting the rhetoric of "progress": the only thing they wanted to uplift was their salary. Still, cinematographers were drawing on the same pool of ideas, using the rhetoric of artistry as shorthand for the rhetoric of class, and drawing exclusionary distinctions between the cultured and the uncultured. Dubray's highbrow rhetoric is an attempt to defuse the lowbrow connotations of Jimmy's authorial persona.[20]

The ASC's public relations battle intensified in 1922, when Foster Goss became the editor of *American Cinematographer*. During his five years in the editor's chair, Goss became a tireless advocate for cinematographers in their quest for professional acceptance. Goss himself was a P.R. man, not a cinematographer, and his paeans to the artistry of cinematography do not necessarily have the ring of sincerity. Had Hollywood costume designers asked him to edit a magazine, Goss would surely have become a tireless advocate for the artistry of costume design. Nevertheless, Goss's editorials offer valuable evidence about the nature of Hollywood norms. Goss was working for a Hollywood institution, writing articles for its own members. Some of his articles address members of other Hollywood institutions that had interests similar to his, from the studios who hired cinematographers to the film critics who reviewed them. Goss's articles may not offer any specialized insight into the mind of the Hollywood cinematographer, but they do offer insights into Hollywood norms, because Goss had to write within the boundaries of those norms to make his arguments convincing. In other words, Goss was a P.R. man selling a product, but he would not be able to sell it if his audience was not willing to listen to his pitch.

If we can agree that Goss's editorials give us some insight into the nature of cinematographic norms, then we may need to revise our understanding of those norms by acknowledging that cinematographers had more openly aesthetic ambitions than is generally acknowledged. Throughout his years as editor of *American Cinematographer*, Goss repeated the cinematographer-as-artist theme over and over again.

Given that identity was at stake, it is not surprising that the cinematographer's job title was itself a point of contention. In 1925 Goss wrote an editorial titled "A Rose by Any Other Name, But—," in which he asked:

If the physician were called medicine man; if the dramatic critic were called paper man; if the surgeon were called bone cutter or meat cutter—what would be the reaction of the members of those respective professions, not to mention the attitude among laymen generally? . . .

We don't know who invented the term cameraman, but we do know that the day when his inventiveness might have been reckoned as genius is dead and obsolete. So should be "cameraman," but it continues to live to give a strictly mechanical—and depreciatory—coloring to a profession to which the rudiments of mechanics is as matter of fact now as are the presses in turning out a book. . . .

The time has come, however, for those who wish to be just and accurate to designate the camera artist as cinematographer.[21]

Like Dubray, Goss hopes to raise the cinematographer into the professional class by downplaying the importance of his mechanical work and affirming the importance of his artistic work.

We can best gauge how important this ideal was to the ASC by looking at a case where it was threatened. In 1926 the academically trained actor Milton Sills gave a speech to the National Board of Review in which he criticized cinematographers for their ignorance of physics and their general lack of culture. This speech was reported in a *Film Mercury* article entitled "Claims Cameramen Are Incompetent." In a stinging reply, Goss argues that cinematographers are cultured in spite of the fact that many lacked academic training: "After all, what shows on the screen does not have to come, in order to meet the most critical artistic standards, through the medium of a university degree or its equivalent—no more than did the works of the master painters have to come from minds 'cultured' according to Mr. Sills' precepts."[22] Goss supports this contention with references to Lincoln, Edison, Steinmetz, Shakespeare, and George M. Cohan. On a more disturbing note, Goss tries to undercut Sills's authority by questioning his sexuality, arguing that "the cinematographer, according to the fairest of present standards, is a pictorialist, judged by what he can produce on the screen, regardless of the fact whether or not he enjoys discussing Freud 'off set' with male stars."[23] The logic (or lack thereof) of Goss's counterattack is not as important as the vehemence with which it is stated, for this vehemence suggests the emotional stake that cinematographers had in their fledgling identities as pictorial artists.

As the examples of Jimmy, Dubray, and Goss suggest, the ASC was attempting to construct a public identity for the cinematographer as a man with a hybrid class identity. On the one hand, the cinematographer usually came from the working class—a background that had supplied him with some admirable virtues: a strong work ethic, a genius for efficiency, and a suspicion of pretension. On the other hand, the elite cinematographer of the ASC had proven his worth by making images that were undeniably artistic. In so doing, he had acquired some of the sophistication of the upper classes. We can call this the mechanics-to-artists narrative. According to this story, the cinematographer retains the admirable virtues of the mechanic, but has adopted the ambitious values of the artist.

The ASC insists that the cinematographer is an artist, but it is not yet clear what sort of art the cinematographer is supposed to make. After all, what is art? Beauty? Imitation? Expression? Art is a famously ambiguous concept, and the ASC made no attempt to resolve this ambiguity. Quite the contrary—the ASC exploited it, suggesting that the cinematographer made many different kinds of art. Because the ASC's goal was to create a convincing new public identity for the cinematographer, it was willing to use any and all evidence that seemed plausible: the cinematographer practiced an art of beauty, an art of realism, an art of expression, an art of skill, and an art of storytelling. Rather than offering a coherent definition of Art, the ASC took advantage of the term's polyvalence. Because there were many different kinds of art, there were many different ways that the cinematographer could prove his artistry.

For instance, when Dubray claims that the cinematographer is an artist, he assumes that art is a matter of creating beauty. In an article entitled "Art vs. Commercialism," he mentions filmmakers who use "a sense of composition to beautify the appearance of pictures."[24] By contrast, when Goss extols the artistry of Hollywood cinematography, he does not use beauty as the primary criterion of value. Instead, he emphasizes technical virtuosity, citing films with spectacular special effects, like *The Ten Commandments* (Cecil B. DeMille, 1923) and *The Thief of Bagdad* (Raoul Walsh, 1924).[25]

For Goss and Dubray, artistry is distinct from storytelling. For others, art is synonymous with storytelling. In 1926 the current ASC president, Daniel Clark, wrote an article claiming that "the ace cinematographer stands as a combination of diversified qualities, including those of the artist, the chemist, the mechanic and the student of human nature."[26] Clark goes on to make two points about the artistry of cinematography. His first point follows Dubray in emphasizing pictorial beauty, but his second point argues that artistic composition works by directing the eye of the spectator to the most important details of the unfolding story. This draws on a long-standing tradition of thinking about the function of composition in narrative pictorial arts, such as painting.

Philip Rosen appeals to the notion of art-as-expression to offer a different way of understanding the artistry of storytelling. In one 1922 article, Rosen writes:

A scene may be photographed in such a manner that it has little or no effect on the audience. On the other hand, the same scene, properly filmed, may conjure in the audience practically any mood that the director wishes to effect. . . .

To the director, as well as the cinematographer, photography should be as the artist's colors and materials are to him. . . .

Every sequence should have its own writing, that is, every sequence should be designed photographically, according to the impression which is designed to be conveyed. The simplest examples of designing for effect are the uses of light for "happiness" and shade for "sorrow."[27]

Rosen emphasizes the artistry of Hollywood lighting by comparing it to painting; this is an early appearance of the "painting with light" metaphor that would soon become so commonplace. For Rosen, artistic lighting is not necessarily beautiful or innovative. Rather, artistic lighting is expressive, enhancing the emotions of the story.

In a different journal, cinematographer Harry Fischbeck used the painting metaphor to link the art of cinematography to another traditional definition of art: art as illusion. He writes, "The artist with the camera, like the artist of the canvas, strives to 'paint' his figures against an appropriate background, making them stand out and convey an illusion of solidity."[28] The illusion is not simply a matter of mechanics. Like a painter, the cinematographer creates an image of a three-dimensional world through his skillful manipulation of light and shade.

Cinematographer Virgil Miller takes the rhetoric of illusionism one step farther. In a 1927 article he bemoans the "disrespect accorded to the cameraman, because of the manual labor that falls to his lot."[29] Drawing on the mechanics-to-artists narrative, Miller argues that the cinematographer should be treated as the equal of the writer or the director. Like Rosen, he argues that the artistic cinematographer is a master of mood lighting, but he mentions one extra benefit: "A cameraman, knowing the desires of the writer, can transfer them to the screen in such a way that the audience lives and laughs and crys [sic], and forgets that they are not the actual beholders of a story's unfoldment."[30] This is something more than a simple illusion of roundness: it is an illusion of presence, giving spectators the sense that they are seeing the fictional world directly.

Here, then, are six different writers offering six different models of the art of cinematography: art-as-beauty (Dubray), art-as-technical-virtuosity (Goss), art-as-clear-storytelling (Clark), art-as-expressive-storytelling (Rosen), art-as-illusion-of-roundness (Fischbeck), and art-as-illusion-of-presence (Miller). It should not surprise us that the discourse of the ASC was so eclectic during its formative years. Prior to the transition to sound, the ASC experimented with several alternate ways of articulating its message about the efficient artistry of the Hollywood cinematographer. There was no need to offer a single consistent vision of the art of cinematography when the ASC could appeal to a wider audience by offering a variety of artistic ideals.

David Bordwell has commented on the role of the ASC within the larger system of classical cinema. He places particular emphasis on the conflict between virtuosic artistry and unobtrusive storytelling:

> The ASC articulated a contradictory task for the cameraman. He was, firstly, to be an artist. The ASC encouraged its members to think of themselves as creative people, comparable to the screenwriter or director. Implicitly, then, each cinematographer's work was to have something distinctive about it. We have already seen the stress laid on individual innovations and virtuosity. But at the same time, the ethos of the craft held that the cinematographer's work must go unnoticed by the layman.[31]

This explanation identifies artistry with innovations and virtuosity, while associating unobtrusiveness with craft. We can occasionally spot moments of flamboyant lighting, but cinematographers usually subordinated their aesthetic aspirations to their professional obligations.

This is already a nuanced account, but I suggest that the discourse of the ASC was even more complicated. As the first institution of its kind, the ASC did not have an obvious model for the articulation of cinematographic discourse. It tried out a variety of ideals: some classical, some not. With respect to the ideal of unobtrusive storytelling, three distinct attitudes appeared. Some argued, in classical fashion, that artistry would sometimes need to be sacrificed to better serve the needs of the story. For instance, in "Art vs. Commercialism," Dubray associates art with beauty, and commercialism with storytelling. Because beautiful photography can distract the spectator's attention from the narrative, Dubray urges his fellow cinematographers to find a compromise: "It is then up to you, cinematographer, to find enough courage to somewhat sacrifice your art, if the sacrifice is to enhance the selling power of the production, as well as it is your duty to display your artistic ability whenever you find an opportunity to do so within the limits of time and expenditure appointed by the producer."[32] The fact that Dubray believes artistry must be sacrificed is directly related to the fact that he defines artistry as pictorial beauty. Beauty arrests the attention, and is therefore an obvious threat to unobtrusive storytelling.

However, there was another way to approach this problem. Producing distracting imagery is not the only way to make a work of art. Philip Rosen makes this position explicit:

> Those things which contribute to the telling of the story in the motion picture are assets; those which do not are liabilities. Anything which distracts the eye and the mind from the story does not belong in the motion picture. . . .
>
> The present-day aim of good photography is to aid the telling of the story by suggesting the mood of the particular scene with the background that is used, by artistically managed lights and shades, and by the compositional use of furniture or scenery.[33]

For Rosen, the art of cinematography is not an art of eye-catching beauty. Instead, it is an art of story-specific decisions about composition and mood. With this definition, the cinematographer does not need to sacrifice his art to the needs of storytelling. His art *is* storytelling—efficient, expressive storytelling. This position is classical, but in a different way. For Dubray, the cinematographer had to choose between the beautiful temptations of art and the more modest craft of unobtrusive storytelling. For Rosen, the cinematographer does not need to make such a choice, because storytelling itself is defined in artful and expressive terms.

The view of Foster Goss represents a third alternative. In the mid-twenties film critics were lavishing praise upon some recent German releases, such as *The Last Laugh* and *Variety*. Goss was outraged that the same critics who ignored the art of Hollywood cinematography were suddenly smitten with the art of German cinematography. Instead of arguing that critics should appreciate the more unobtrusive virtues of Hollywood films, Goss argued that Hollywood should fight spectacle with spectacle. In one editorial, he writes:

> Whatever may be the excellencies or the crudities of the German-made motion pictures, they at least are centering attention on one long-neglected fact—that the cinema is an art distinct and complete in itself. However inanely simple such a statement may seem to be, it is still true that pictures are largely literature, paintings, etc., as expressed in cinematography. It's been a case of "the play's the thing" rather than "the picture's the thing."[34]

Hollywood cinematography may be unobtrusive, but not if Goss can help it. Unlike Rosen, Goss believes that pictorial artistry is at odds with unobtrusive storytelling. Unlike Dubray, Goss refuses to take the classical step of subordinating style to story. Significantly, the title to Goss's editorial is "Subservient Art?" The question mark registers Goss's unwillingness to assume a subordinate role.

Although Goss provides evidence for a nonclassical account of the Hollywood style, we should not go so far as to say that Goss rejects storytelling altogether. In the next paragraph, he writes, "Simple stories, deliberately told, attain a forcefulness which indicates what is still to come in the cinema art." This suggests that Goss is departing from Dubray and Rosen in a very particular way. Dubray and Rosen both assume that storytelling, at its best, should be unobtrusive. Goss does not make this assumption. He is in favor of storytelling, but he sees no reason to believe that a storytelling style is necessarily unobtrusive. For Goss, narrative and spectacle are fully compatible values.

The point here is not that Goss was secretly subversive, encouraging cinematographers to defy the norms of the invisible style. Rather, and perhaps even more striking: if Goss's statements seem to be outside the norms, this might be because we need to revise our understanding of the norms themselves by acknowledging that they were more eclectic than we usually suppose, at least during the formative years of the ASC. During the 1920s *American Cinematographer* did more than any other publication to articulate the norms of Hollywood cinematography. From 1922 to 1927 this journal was edited by a man who believed that cinematographic imagery at its best was designed to be noticed. Should we dismiss this detail, on the grounds that Goss was just a P.R. man looking for a flashy product to sell? Perhaps, but I would suggest an alternative possibility: the goal of unobtrusiveness was one ideal among many within the systematically eclectic discourse of the ASC.

Admittedly, the ideals of the ASC did not remain eclectic forever. Its cinematographers eventually found that some rhetorical strategies were more effective than others. The discourse grew more homogeneous after the transition to sound, drawing cinematographers closer to the classical norm. Still, the fact that classical modes like the "invisible style" and the "illusion of presence" were commonplaces in later years should not lead us to conclude that their use was an unquestioned assumption during the twenties. Given the ambitious agenda of the ASC, it makes more sense to suppose that this young institution adopted an experimental strategy—a strategy of testing out different rhetorical options and advancing a range of aesthetic ideals, some more classical than others.

Of course, discourse is not the same thing as practice. How should the preceding discussion affect our approach to the study of style? It is tempting to say that we should start by looking for extravagant displays of pictorial spectacle, to confirm the idea that the "invisible style" was not yet an inviolable norm. To be sure, such displays are not difficult to find: elite cinematographers like Charles Rosher, John F. Seitz, and Arthur Edeson crafted images of remarkable beauty throughout the silent period. However, we should remember that beauty was just one aesthetic ideal among many. Most of the aesthetic ideals of the ASC were fully compatible with the goal of storytelling. Without grasping this crucial point, we will not be able to understand why the ASC eventually did endorse the ethic of invisibility after the transition to sound. The ASC endorsed this ideal not because it was abandoning its claims of artistry, but because its members converged on the idea that the cinematographer could be an artist and a storyteller at the same time. Instead of looking at extravagant displays of spectacle, the next few chapters will attempt to explain a more important process: the institutionalization of artistry.

A skeptic might say that the concept of an artistic convention is itself paradoxical. Don't conventions produce a style that is neutral, zero-degree, and artless? Not necessarily. First, Hollywood films do not all look the same, because most lighting conventions were conventions of *differentiation*. A man should not be lit like a woman, a crime film should not be lit like a romance, and a back alley should not be lit like a ballroom. Learning how to light involved learning how to light differently.

Second, the ASC's notion of artfulness was centered not on the ideal of originality, but on the ideal of functionality. A technique is not artful by virtue of being innovative. It is artful because it accomplishes a recognizable artistic function, such as establishing mood, enhancing characterization, adding realistic detail, or creating pictorial beauty. For this reason, the ASC was not interested in encouraging its members to innovate at all costs. Instead, the ASC encouraged its members to adopt time-honored techniques for establishing mood, enhancing characterization, and so on. In other words, the ASC hoped to *increase* the conventionality of Hollywood cinematography. If a technique was effective, why shouldn't everyone use it? A cinematographer who had to wait for inspiration was a handicap; a cinematographer who knew a wide range of experience-tested conventions was the perfect artist for a commercial industry based on mass production.

This brings us back to the wise observations of Jimmy the Assistant: the ideal cinematographer is the one who combines the virtues of efficiency with the values of art. The next three chapters will explain how cinematographers produced that combination.

2 From the Portrait to the Close-Up

The American Society of Cinematographers wanted to create a public identity for the cinematographer as a skilled professional artist. The field of portrait photography provided the ASC with a useful model, in part because so many of the organization's earliest members began their careers as portraitists, including Charles Rosher, Arthur Edeson, Tony Gaudio, John Leezer, and Karl Struss. Hollywood cinematographers quickly surpassed the portraitists in terms of technical resources, but the discourse and practice of portraiture would leave a lasting impression on the field of cinematography by defining the task of figure-lighting as an art of characterization and illusionism.[1]

We can start by taking a close look at the conventions of portrait photography. Arthur Hammond's 1917 article in *American Photography*, entitled "Home Portraiture," offers a good introduction to the field. According to Hammond, the amateur and the professional needed but two things to light a portrait: a window and a reflector. The article includes several diagrams showing the right way and the wrong way to use these two tools. The diagram illustrating the "right" way appears in figure 2.1. By following this diagram and placing the camera at spot G, the portraitist can create what Hammond calls "ordinary lighting": "One side of the face is fully lighted, and on the other side, the side away from the window, there is a triangle of light on the cheek, just below the eye."[2] By carefully adjusting the window shades, the photographer can modify the lighting and control the highlights. Hammond has remarkably precise instructions about the placement of the highlights: "With good lighting there should be a highlight on the forehead, just above the eye on the lighted side of the face, and also on the bridge of the nose, on the tip of the nose, on the lips and chin, and also a bright spot in each eye."[3] Figure 2.2, a self-portrait of photographer J. B. Wellington, is a good example of the technique. All the highlights are in the proper place: the forehead, the bridge of the nose, the tip of the nose, and the bright spot in both eyes. No doubt there would also be highlights on the lips and chin, if not for Wellington's mustache.

In spite of his precise stipulations about highlights, Hammond does allow some room for variation. For instance, the presence of a bay window might allow the photographer to light a portrait without using a reflector. Alternatively, a photographer could try out a technique called "Rembrandt lighting." The technique employs the same basic "ordinary lighting" setup. Instead of changing the lighting, the photographer simply moves the camera. In figure 2.1, the "H" represents the ideal spot for "Rembrandt lighting."

"Rembrandt lightings" had been around in photography for decades. In his 1891 book *The Studio, and What to Do in It*, Henry Peach Robinson explains: "The Rembrandt portrait is usually a head, more or less in profile, lighted from behind and the side, and as unlike anything Rembrandt ever painted as possible. I have always objected to this title for these

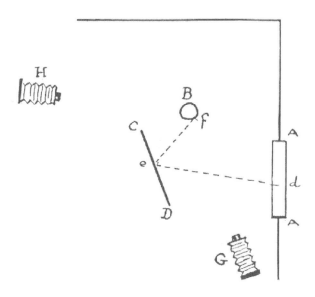

Figure 2.1 The subject (B) is lit by the window (A). The reflector (C–D) helps fill in the shadows. Placing the camera at G produces ordinary lighting. Moving the camera to H produces "Rembrandt lighting."

FROM ARTHUR HAMMOND'S ARTICLE "HOME PORTRAITURE," IN *AMERICAN PHOTOGRAPHY* (MARCH 1917).

Figure 2.2 A self-portrait of the photographer J. B. Wellington.

FROM *AMERICAN PHOTOGRAPHY* (1917).

shadow pictures, but the name sticks, and I accept it."[4] Robinson is right to point out that "Rembrandt lighting" is a potentially misleading term. Although the Dutch master Rembrandt van Rijn did occasionally light from the side-back position, he favored a soft light from above. Nevertheless, photographers, like their peers in cinematography, were often struggling to justify their work as art, and the "Rembrandt" tag served to foreground their artful manipulation of daylight.[5]

Not all photographers at the time relied on window lighting for their portraits. Some professional photographers had invested in artificial lights. This technical change did not involve a shift in aesthetic strategies. Photographers liked the reliability of artificial lights, but they preferred the natural look of daylight. Indeed, the ideal of "naturalness" carried a certain rhetorical force. Artificial lights were convenient, but they were to be used in a way that duplicated the inherently superior light of nature. In the 1918 edition of the *American Annual of Photography*, studio photographer T. W. Kilmer describes the professional's dilemma:

> For many years man has been striving to produce a light that will take the place of daylight as an illuminant in photographic portraiture. He has had a hard job. Daylight is certainly in a class all by itself when it comes to using it for this purpose. Its softness, its subtleness, its actinic quality, its broadness, its various moods, all made it the ideal illuminant. Although ideal in character,

it is nevertheless difficult to master, for one moment it lights your subject with a full blaze of bright light, only to be followed by a period of soft, dull light caused by a cloud scudding across the sun.[6]

Kilmer goes on to suggest a simple solution: use artificial lights to duplicate the soft look of daylight. This desire for softness causes Kilmer to reject carbon arc lights, which give off light from a point source. Instead, Kilmer recommends using Cooper-Hewitt lamps, which produce a softer light because they illuminate from a larger area. Kilmer also recommends supplementing this light with a few nitrogen lamps. As for aesthetic matters, Kilmer simply advises the photographer to set up the lights so that they reproduce the established style: "Turn on your Cooper-Hewitts, and then add 1, 2, 3, 4 or more 100 watt nitrogen lamps to your Cooper-Hewitt illumination until your subject looks as though he was lighted by daylight."[7] Satisfied with the look of daylight, Kilmer recommends using artificial lights to duplicate that look efficiently.

Some photographers figured out ways to overcome the hardness of the arc light. In a 1913 article in the *British Journal of Photography*, George F. Greenfield describes lighting with an enclosed arc light.[8] Like Kilmer, Greenfield is wary of the light's hardness, but he solves the problem by using a curved reflector to bounce light toward the subject, thereby using artificial lights to recreate the softer look of daylight. The soft key comes from a front-side position, while an additional reflector provides some fill for the shadow side. Arc lights could also be used to create "Rembrandt lighting," as discussed in a 1916 article by George R. Henderson. Again, the photographer avoids hard light, preferring to soften the light by bouncing it off an umbrella.[9]

In short, whether British or American, amateur or professional, the typical portrait photographer of the time favored the soft look of daylight, so much so that even artificially lit portraits were lit in this fashion. At first glance, the aesthetic justification for this approach seems straightforward enough: this lighting provides the modeling necessary to create a plausible likeness of the sitter. Photographers were concerned to create a certain kind of realism—a realism defined as the "illusion of roundness." Arthur Hammond writes:

> Our vision is stereoscopic because the two images seen by the two eyes are merged and coalesced into one image, just as the two pictures in a stereoscopic photograph are seen as one. This gives us a sense of roundness and relief which, in a photograph made with only one lens we must suggest by

means of varying intensities of light, halftones and shadows. . . . This is why lighting in a portrait is so important. The direction of the lighting determines the positions of the highlights, halftones and shadows on the face and these indicate the shape of the features and consequently the likeness to the individual portrayed.[10]

According to Hammond, the photographer must compensate for the two-dimensional nature of the medium by using light to create an illusory sense of depth. The photographer creates a likeness by molding the contours of the shape of the sitter's face.

Many of Hammond's technical recommendations contribute to the goal of creating a sense of depth. For instance, Hammond insists, "To obtain the maximum relief and roundness, the light should come from one source only." Hammond also advises stretching sheets of muslin over the window to soften the light, if necessary.[11] Because a soft light creates multiple degrees of gradation between the light side and the shadow side, it can enhance the sense of depth.

Notice that Hammond does not suggest that the camera's automatic use of perspective will supply the required sense of depth. The mechanical camera does not capture depth; the individual photographer creates depth, through the skillful manipulation of light and shade. Likeness is created, not captured. This rhetoric emphasizes the creative contributions of the individual photographer.

Photographers who were concerned to justify photography's status as an art would take this argument about the photographer's individual contributions even farther. Antony Guest, in his 1907 book *Art and the Camera*, writes:

> What is likeness? It is not altogether an objective matter that makes its appeal by a set arrangement of form and colour, so that what one sees every one else sees, and as to which there is no room left for diversity of opinion. It is at least to an equal extent subjective, depending on the point of view of the beholder, his mental attitude, and the degree of sympathy that he feels for the person portrayed. . . .
>
> Hence it seems that some other aim needs to be substituted for that of mere similitude, or, at least it should be supplemented by something more trustworthy; and that which naturally suggests itself is character.[12]

Guest opposes the weak notion of "mere similitude" to the stronger goal of capturing "character." According to this theory, appearances may change from moment to moment, but each subject has a more-or-less consistent character that can be represented by the artistic photographer. Again, the work of Rembrandt provides the portrait photographer with a model. If Rembrandt were merely a virtuoso with light and shade, he would be a minor figure. Traditionally, what makes Rembrandt a great master is his legendary ability to peer into the souls of his subjects.

Although "character" is invoked as a way to avoid troubling questions about what constitutes a "likeness," the concept of "character" is itself far from simple. The photographers' discourse of "character" is almost invariably complicated by a discourse of sexual difference, as many photographers start with the assumption that men and women have different degrees of "character." This ideology of difference impacts their formal choices in concrete and specific ways. In his 1919 book *The Fine Art of Photography*, pictorialist and portraitist Paul L. Anderson offers a detailed discussion of the way gender and character impact photographic technique:

> Men are most likely to have strongly marked characters, since their mode of life tends to develop the mental processes and to encourage decision, whereas our present unfortunate ideals of feminine beauty incline toward mere regularity of outline and delicacy of complexion. One finds, nevertheless, a good many women whose features express mental activity and firmness of will, the higher beauties of the mind rather than the mental indolence which is imperative in the cultivation of what is popularly termed beauty.[13]

According to this theory, character is visible on a person's face to a greater or lesser extent. A person with a strongly marked character will have more lines on the face, a result of mental exertion; conversely, a person with less clearly marked character will have a more smooth complexion. Furthermore, character (both as an internal state and as its external manifestation) supposedly has a tendency to vary with respect to men and women. To his credit, Anderson refuses to naturalize this distinction, arguing that "the present variation seems to be rather the result of education and training than of anything else." Nevertheless, he resigns himself to the distinction, noting that "for the present the facts are as stated."[14]

Ultimately committed to the "facts as they are," Anderson uses his generalizations about character to support various proposals about photographic conventions. He divides his recommendations along lines of gender and age, writing:

We are accustomed to associate brightness and vivacity with children, and these qualities are suggested by a high-keyed print, transparent and full of light. . . . To a less extent the same is true of portraits of women, though here the scale may be extended, more contrast being used, even (in the case of women of strong character) approaching the full-scale, powerful effects which are valuable in portraying men. Evidently, men less accustomed to commanding positions, that is, artists, writers, students and the like, approach more nearly to the feminine gentleness of character, and they, since their work is more in the realm of the imagination, are generally to be rendered with less contrast and vigor than those who have charge of large affairs.[15]

Anderson makes gender-specific (and age-specific) recommendations about the formal features of tonality and contrast. Specifically, a portrait of a man, particularly a powerful one, should have strong contrasts of light and dark. Although some portraits of women can approach this range, the general rule is to employ a brighter overall tonality with images of women. He offers two of his own pictures as an example (see figs. 2.3 and 2.4). The portrait of the man has strong contrasts of light and dark, with darkness providing the dominant tonality. The portrait of the woman has much less contrast, and brightness is the dominant tone.

Anderson does not explain why the control of contrast and tonality serves to differentiate character. However, it seems likely that the technique works in two ways: emphasis and expression. According to Anderson, people with strong characters have stronger lines in their faces. High-contrast lighting and printing would serve to make these lines more visible, while low-contrast lighting would emphasize the smoothness of a sitter's face. Applying this logic, to emphasize a sitter's "decisive" character, a photographer might use contrasty chiaroscuro to bring out the hard-wrought lines on a person's face. To emphasize a sitter's "indolent" cultivation of beauty, a photographer might use less contrast to emphasize the smoothness of a person's complexion.

Other photographers suggest different ways of emphasizing character. For instance, Arthur Hammond recommends varying the angle and quality of the light: "The lighting should always be suited to the subject, a strong shadow lighting might be appropriate for a man, but a softer, flatter lighting would usually be better for a young woman or a child."[16] Hammond does not share Anderson's willingness to interrogate cultural assumptions about gender and "character," and he does not explain the justification for his recommendations; however, it seems likely that this is another case of differentiation-through-emphasis. A soft, flat (i.e., fron-

Figure 2.3 Strong contrasts and dark tonalities in a masculine portrait by Paul L. Anderson, in his 1919 book *The Fine Art of Photography.*

Figure 2.4 Anderson uses gentler contrasts and lighter tonalities in a feminine portrait, also from *The Fine Art of Photography.*

tal) light works to eliminate the wrinkles that may be more apparent with side-lighting. It would be counterproductive to use this technique on a man, since emphasizing facial lines is a way of emphasizing character.

A strategy of emphasis can be complemented by a strategy of expression. According to Anderson, a high-keyed print suggests the qualities of "brightness" and "vivacity." The formal elements carry certain associations, and these associations can be used strategically to express the sitter's personal qualities. In a chapter on "Appropriate Treatment," Antony Guest writes:

> It is often overlooked that additional expressiveness is to be obtained through the decorative influence if applied with discrimination. Pictures of people are sometimes composed as if the beauty of lines and masses were a thing apart, a sort of gratuitous adornment in no way relating to the personality portrayed. To instance a ridiculous extreme, we may suppose a portrait of Lord Kitchener treated with delicacy, while that, say, of a pretty actress is composed with severity of line and an impressive chiaroscuro.[17]

Guest's point is that the reverse treatment would be more appropriate. Although Guest does not make his assumptions about gender explicit, it is no surprise that his recommendations follow the basic strategy of differentiation outlined by Anderson. The commanding male figure should be represented with "severe" lines and "impressive chiaroscuro," while the pretty actress should be treated with more "delicacy." Extending this expressive analysis to Hammond's recommendations, we find that he prefers "strong" lighting for men, and "soft" lighting for women.

Some photographers recommended strategies for manipulating the psychological state of the sitter. For instance, one article in *Photographic Review* advises photographers to argue with their male sitters, in order to bring out the lines of tension in their faces. By contrast, photographers should put women at ease by having conversations about children, music, and flowers. The article even suggests placing flowers nearby as pretexts for starting such conversations.[18]

In short, a discourse of gender difference intersects with the photographers' discourse of character. The intersection yields practical results, in the form of specific conventions, as follows.

1. *Direction of light.* Frontal lighting smooths wrinkles, while side- and top-lightings emphasize them. The former is preferred for women, while the latter is preferred for men.

2. *Quality of the light.* Diffusing or bouncing the light softens the edges of the shadows. Cooper-Hewitts also produce soft-edged shadows. Undiffused arc lights produce hard, sharp shadows. Some photographers are opposed to the use of hard lights, regardless of the subject. In general, softness is preferred for images of women for two reasons: the expressive associations of the term "softness," and the tendency for soft shadows to deemphasize facial lines.

2. *Contrast* (lighting). Even a soft frontal light will cast some shadows, but weakening the contrast by adding a strong fill light will make those shadows even less salient, emphasizing the smoothness of the sitter's features. Decreasing or eliminating the fill light will emphasize shadows (and therefore character), particularly when combined with hard side-lighting. On an expressive level, pictures of women feature "gentle" gradations in tone, while pictures of men feature "strong" shadows.

4. *Overall tonality.* Overexposing a woman's face is one way to smooth out lines. Photographers prefer darker tones for images of men, to create a mood of "seriousness."

5. *Lens diffusion.* Using soft lenses or placing gauze over a lens softens the image, with predictable results to expression and emphasis.

6. *Lens focus.* Even when not working with a specially designed "soft" lens, a photographer could choose to throw a woman's face out of focus, thereby smoothing out lines. Used in another way, the technique can also add character to a picture of a man: by softening other areas of the frame, a photographer could draw attention to a sharply focused, well-lined face.

7. *Retouching.* This is another way a photographer could soften all or part of a picture and smooth out the wrinkles in someone's face. Some photographers preferred this option; others preferred to manipulate the lens focus instead.[19]

8. *Contrast* (developing and printing). Yet another way to influence the contrast of the image is to use different developing or printing techniques. For instance, overexposing and underdeveloping the negative will result in a low-contrast print, emphasizing soft gradations over strong contrasts. Meanwhile, underexposing and overdeveloping the negative will result in stronger contrasts. Similarly, printing techniques could be combined with exposure and/or developing techniques to adjust the contrast of an image.

These conventions can intersect with other conventions, such as the aestheticizing conventions of soft-focus pictorialism. Soon after *Vanity Fair* begain publishing in 1914, pictorialist Arnold Genthe made several portraits for the magazine (see figs. 2.5 and 2.6). Regardless of the subject, most of Genthe's portraits are in the softened style that was a trademark of pictorialism. Here, Genthe's aesthetic approach has prevented him from using lens diffusion as a tool for the emphasis and expression of "character." Still, Genthe manages to introduce distinctions with other techniques, such as the angle of the light. His portrait of Paderewski (fig. 2.5) uses a top-side-light to emphasize the complex shape of the sitter's face. The forehead has several planes, the nose stands out from the cheek, and the cheek itself is an intersection of multiple planes. The portrait of Lucile Cavanaugh (fig. 2.6) uses a similar lighting scheme, but it has been adjusted to a top-front-side position, allowing the area from the forehead to the cheek to the chin to be rendered as one smooth plane.

Compare Genthe's work to the work of Victor Georg, another *Vanity Fair* regular. Like Genthe, Georg tends to soften all his portraits. However, Georg is less likely to move his key-light to construct difference; instead, Georg uses contrast and tonality. Look at the shadows under Castle's nose,

Figure 2.5 The top-light models the features of Ignace Paderewski, in a portrait by Arnold Genthe.

FROM *VANITY FAIR* (SEPT. 1918).

Figure 2.6 Genthe uses gentler modeling in his portrait of Lucile Cavanaugh.

FROM *VANITY FAIR* (JULY 1917).

lip, and chin (fig. 2.8), and compare them to the corresponding shadows on Arliss (fig. 2.7): Arliss's shadows are much darker. Indeed, the overall tonality of the Arliss portrait is darker. Georg uses tonality to express Arliss's "serious" character, and he uses contrast to give emphasis to the external signs of that character—the lines on Arliss's face. In comparison, the handling of Castle does not reveal character. Her portrait flattens into a decorative pattern: the curved hands at the bottom of the frame mirror the curved feathers at the top, while the curve of her hat mirrors the curve of her shoulder.

Not all of the photographs of Genthe and Georg operate at these extremes, but a binary strategy seemed to operate as a norm, a background against which departures could be measured. There was at least one photographer who departed from this norm in a more significant way: Baron Adolph De Meyer, a European aesthete who was almost certainly not a real baron. De Meyer rejected the rhetoric of naturalness and developed a flamboyantly artificial lighting style. In figure 2.9, the backlight outlines John Barrymore's famous profile, but the space between the nose and the

Figure 2.7 George Arliss emerges from a dark background in a portrait by *Vanity Fair* photographer Victor Georg (March 1916).

Figure 2.8 A carefully composed portrait of Irene Castle, by Victor Georg.

FROM *VANITY FAIR* (JAN. 1916).

ear looks almost flat. Even the profile's outline is not nearly as clean as it could be. Had he wanted to represent the profile more clearly, De Meyer could have outlined it by placing the backlight at eye level. Instead, De Meyer chose to employ the low backlight. The result is an unpredictable, nearly abstract composition, as some of the profile's features pop out, while other features blend into the background.[20]

Rather than imitating the natural look of daylight, De Meyer relies almost exclusively on artificial lights, often in a three-point setup. During the late teens, Hollywood was also using the three-point setup more and more often, though it is not clear whether De Meyer influenced Hollywood, Hollywood influenced De Meyer, or both. This similarity of technique does not imply a similarity of approach. With its commitment to narrative, Hollywood does not follow De Meyer's practice of turning every portrait into an abstract picture. Rather, Hollywood uses lighting to capture and create varying degrees of character, be it the character of fictional individuals or the character of the stars playing the roles. This commitment to character led Hollywood to adopt a set of norms that were, if not

Figure 2.9 Artificial lighting in a portrait of John Barrymore, by Adolph De Meyer.

FROM *VANITY FAIR* (JAN. 1920).

identical to those of the photographers, structurally analogous to them, emphasizing and expressing different degrees of character for men and women.

Individual techniques were transformed in the transition from still photography to cinematography, in spite of the fact that many of the cinematographers who helped develop the Hollywood style had experience in portraiture. One reason is technological: the motion picture industry moved much more quickly to the use of artificial lights than the still photography industry did. For instance, a 1917 article in the photography publication *Studio Light* describes the Campbell Studio, located on New York's Fifth Avenue. The sophisticated location is a sign that the studio was at the top of the profession, but the article notes that the studio's primary light source is a window, and that most of the accompanying portraits are in the window-and-reflector tradition.[21] Even Edward Steichen, one of the most successful photographers in the world, claims in his autobiography that he did not begin to use artificial lights regularly until 1923, when he took a job at *Vanity Fair*.[22] By contrast, in 1917 more and more film studios were painting over their glass studios and relying exclusively on artificial lights, in part because the fickleness of daylight inevitably produced errors in continuity.[23] Even photographers who did use artificial lights had a different technological base than their Hollywood counterparts. With longer exposure times, photographers did not need as much light. *Studio Light* reports that an exposure time of one or two seconds was typical for a 1914 portrait photographer. By contrast, a cinematographer running his camera at 16 frames per second worked with an exposure time of approximately 1/32 second. (This assumes a 180-degree shutter. There were other shutters in use, but this will do as an approximation.) Because of this difference in exposure time, the cinematographer needed more instruments—and stronger instruments—than did his peers in the photographic profession. This may also explain why cinematographers did not follow Greenfield's advice and bounce their arc lights off reflectors to create more softness: too much light is lost in the process. Although cinematographers do use lamp diffusion, their key-lights are still much harder than those of still photographers. The links between figure-lighting and portraiture do not exist at the technical level; they exist at a structural level, in Hollywood's gradual adoption of portraiture's principles of differentiation.

Figure 2.10 comes from Maurice Tourneur's *The Wishing Ring* (1914), a film made before the wholesale switch to artificial lights.[24] Daylight probably supplies the overall light here, though the hard shadow on the subject's right cheek suggests that an artificial light is supplementing the key. Tourneur and his cinematographer John van den Broek were famous for their pictorial skills, and this image shows the same attention to modeling that was so prized by still photographers. The difference in brightness on the two sides of the face lend it a sense of roundness, while costume and set design provide the necessary separation of foreground and background.

In comparison to later films, the films of this period are much less rigorous about the norms of sexual difference. This is not because Hollywood filmmakers had a different ideology. In order to signify difference of any kind (male/female, day/night, etc.), it is necessary to have a range of options, and in 1914 the still photographer simply had more options than the narrative filmmaker. Some of those options were technologically impossible for the cinematographer (e.g., retouching); some options were not yet available (e.g., soft lenses); and some options were available but used for other tasks (e.g., contrast and tonality might be used to mark time of day). In the late teens, the palette of options would expand, as

Figure 2.10 Portrait lighting in Maurice Tourneur's *The Wishing Ring* (1914).

would the degree of differentiation in the norms of photographing men and women. Cinematographers would never have the option of retouching, but they began to use a variety of softening techniques, and they soon had more artificial lights than all the portrait photographers in New York. The greater flexibility allowed cinematographers to construct all sorts of distinctions, including distinctions between close-ups of men and close-ups of women.

As Barry Salt and Kristin Thompson have argued, the adoption of softening techniques is a case where the influence of still photography is remarkably direct, on the level of both technology and technique.[25] John Leezer is often credited as the first cinematographer to use a soft-focus lens, while filming a 1916 film called *The Marriage of Molly-O*.[26] A former portrait photographer, Leezer used a lens designed by Paul Anderson's colleague, photographer Karl Struss (who later became a cinematographer). The motivation behind the use of the lens was unabashedly aesthetic: Leezer's studio even gave a bonus for artistic work. A few years later, D. W. Griffith, one of Leezer's bosses at Triangle, hired portrait photographer Hendrik Sartov for the specific purpose of applying his soft-focus style to the close-ups of Lillian Gish. Griffith and his crew strove to combine the function of creating pictorial beauty with the function of enhancing characterization. Films like *Way Down East* (1920) routinely use much more lens diffusion on close-ups of women than on close-ups of men, as in figure 2.11, a heavily diffused close-up of Gish. Soon some female stars began to request heavy diffusion on all their shots.[27]

Sartov also contributed to Billy Bitzer's ongoing experimentation with backlighting. In his autobiography, Bitzer claims that he discovered backlighting by accident when he noticed the sun shining on Mary Pickford

Figure 2.11 Heavy lens diffusion and a strong top-backlight in *Way Down East* (1920).

and Owen Moore during a lunch break.[28] Whether this story is true or not, backlighting became Bitzer's default norm for exterior scenes. A few years later Bitzer began using artificial lights to achieve the same effect. Bitzer's assistant Karl Brown describes the technique, treating it as a gender-specific convention: "You could also burn [a spotlight] into the back of a frizzy-headed blonde and make her look like a haystack afire."[29] Whereas De Meyer happily aestheticized all his portraits, Bitzer was more likely to reserve the backlight for women, even at the risk of continuity violations. In the image from *Way Down East*, note how the hyper-aestheticized Gish blends almost seamlessly into the pattern behind her.

Backlighting is sometimes cited as the primary example of a gendered stylistic code, but it can actually serve a wide range of purposes.[30] In Cecil B. DeMille's 1917 film *A Romance of the Redwoods*, cinematographer Alvin Wyckoff uses backlighting selectively, favoring Mary Pickford (Jenny), rather than co-star Elliott Dexter ("Black" Brown). In the style typical of a 1917 film, both characters are first shown in special introductory shots. In Pickford's shot, not only does the backlighting emphasize her famous blonde hair (which would have looked dark on orthochromatic film); it also serves as a metaphor for Jenny's angelic morality. Meanwhile, Dexter receives a harsh light that produces unpleasant shadows on his face. The rest of the film continues this pattern, using backlighting to characterize Jenny as an angelic figure. The stylistic exceptions actually support this narrative pattern. When Jenny is threatened, her halo disappears. When "Black" Brown experiences a moment of redemption, a backlight symbolically marks his shift toward Jenny's values.

While this example generally confirms the point that backlighting was originally associated with women, cinematographers soon began to use the technique on all subjects, regardless of gender. In Pickford's *Suds*

(John Francis Dillon, 1920), cinematographer Charles Rosher backlights everything in sight. He backlights Pickford, even though her character is unglamorous. He also backlights the male lead and various supporting characters. He even backlights a horse. The function of characterization has been subordinated to another function: securing an illusion of roundness. As cinematographers realized that backlighting was an efficient way to separate foreground from background, they began to use it as a default convention for almost all situations.

Backlight soon became a default norm, but a cinematographer could still intensify its use to produce a special effect. In *Suds*, for instance, when Pickford's unglamorous laundry worker imagines that she is a duchess, Rosher adds strong diffusion to create a deliberately stylized image of noble glamour. This helps to explain why backlight seems to be associated with women, even though it is used for men as well. With women, backlight is often combined with lens diffusion, which intensifies the "halo" effect by making the backlight glow.[31]

Though backlight is easier to notice, the handling of key and fill is equally significant. In this period of experimentation during the late 1910s and early 1920s, cinematographers tried out various options. For instance, Joseph August often lights William S. Hart in a way that maximizes the "ruggedness" of the star's features, as in figure 2.12, from *The Toll Gate* (Lambert Hillyer, 1920). The key-light creates a strong nose shadow (which is further emphasized by the lack of a fill light), while a kicker emphasizes his cheekbones and adds an additional accent to the nose. The "strong" contrasts carry the "masculine" expressive connotations that portrait photographers prized.

Contrast this with Charles Rosher's handling of Mary Pickford in *Daddy-Long-Legs* (Marshall Neilan, 1919) in figure 2.13. As Barry Salt has pointed out, this close-up is lit with at least four different instruments: an arc to the right of the camera, an arc to the left, and two arcs for backlight.[32] Rosher has not yet adopted the "soft style": there is no lens diffusion, and he lights Pickford with multiple hard lights. Still, he manages to "flatten" Pickford's face by using two keys of equal intensity, thereby eliminating nose shadows. There is no point in claiming that one of these lights is the key and one is the fill. Rosher is clearly trying to use two key-lights of equal intensity, to ensure that each light cancels out the shadows made by the other. This "double-key" technique draws attention to her eyes and mouth. Meanwhile, the two backlights prevent the entire image from going flat, enhancing depth by emphasizing the famous curls in Pickford's hair. The result is a careful balancing of flatness and depth,

Figure 2.12 Rugged lighting for William S. Hart in *The Toll Gate* (1920).

producing a moderate illusion of roundness while smoothing the features of the subject. Rosher had experience in portraiture, but he was not simply imitating an established technique. The British photographer who had once been Rosher's employer would later marvel at Rosher's innovations, wondering if his own studio had the technical facilities to duplicate them.[33]

Both of these images can be seen as attempts to apply the differentiating logic of portraiture to the art of the close-up. Still, many important cinematographers were experimenting with styles that made no strong distinction between men and women. For instance, former portrait photographer Arthur Edeson uses a fairly simple strategy in *Robin Hood* (Allan Dwan, 1922)—a key from the side, with a generous amount of fill. The technique produces similar modeling for both subjects, whether male or female, establishing an illusion of roundness without being used as a tool for differentiation.

John F. Seitz provides a more complicated example. Working with director Rex Ingram, Seitz experimented with a style that Barry Salt has called "core-lighting." This strategy employs two side-lights, leaving shadows in the middle of the subject's face.[34] Figure 2.14 provides an example, from the 1922 film *The Prisoner of Zenda*. Ingram was a former sculptor, and he was explicitly interested in giving his images a quality of three-dimensionality. In one 1922 article he wrote: "It is modeling obtained by judicious arrangement of light and shade, that enables us to give something of a stereoscopic quality to the soft, mellow-toned close-up, which takes the place of the human voice on the screen and helps to make the audience as intimate with the characters as if they had known and seen them constantly in everyday life."[35] The core-lighting technique draws on the discourse and practice of photography, but it is also something of a

Figure 2.13 Smooth lighting for Mary Pickford in *Daddy-Long-Legs* (1919).

departure. Like Hammond, Ingram insists that the illusion of roundness is a matter not of mechanical reproduction but of the skillful handling of light and shade. Unlike Hammond, however, Ingram pushes the "illusion of roundness" concept so far that it begins to resemble another concept: the illusion of presence. The audience does not simply see depth in the image; the audience actually imagines being in contact with the subject, coexisting in the same space. Perhaps Seitz and Ingram were reasoning that two side-lights would produce even more depth than one. If so, they were departing from the discourse and practice of portraiture. The double-side-light was not a popular technique, because it fell short of the "natural" ideal. In particular, it was not an appropriate technique for portraits of women, since these portraits were supposed to emphasize smoothness, rather than character. Such aggressive core-lighting was usually reserved for strong men, as in Edward Steichen's 1908 portrait of Theodore Roosevelt.[36] This is what makes figure 2.14 so unusual: according to the norms of portraiture, Alice Terry is overmodeled; Seitz and Ingram emphasize her features too strongly.

The most daring cinematographers would continue their experiments throughout the twenties. In the late 1920s Seitz and Victor Milner were both moving their key-lights closer to a top-light position, partly because they wanted to imitate the original lighting technique of Rembrandt. Meanwhile, cinematographers like Rosher and Karl Struss were exploring the possibilities of the soft style, using lens diffusion and lamp diffusion to mimic the soft look of pictorial photography. Still, this experimentation took place within the context of increasing stability. Options like core-lighting and double key-lighting were becoming even more unusual, as three-point lighting had become the unquestioned norm, even among elite D.P.s like Seitz and Rosher. For instance, compare Rosher's work in

Figure 2.14 Two cross-lights provide strong modeling in *The Prisoner of Zenda* (1922).

figure 2.15 (from Sam Taylor's 1927 comedy *My Best Girl*) to his earlier work, in figure 2.13. Whereas the image from *Daddy-Long-Legs* is lit with two keys, the image from *My Best Girl* is much closer to the three-point norm, with a gentle but noticeable difference in intensity between the key and fill.

Significantly, Rosher has employed at least three kinds of softness in crafting his 1927 portrait of a female star. First, the lens produces softer focus. Second, all of the lamps have diffusion, softening the edges of the shadows. Third, the image as a whole has less contrast, as Rosher has composed it with light shades of gray throughout. Still, a functional similarity stands out. In *Daddy-Long-Legs*, Rosher used two keys to smooth out Pickford's features, according to the theory that the cinematographer should work to reduce the lines of character on a woman's face. In *My Best Girl* Rosher has changed his technique, but his goal is still the same: to photograph Pickford according to a gender-specific strategy of representation.

Figure 2.15 Soft modeling on Mary Pickford in *My Best Girl* (1927).

Exteriors presented a distinct set of problems for the cinematographer who wanted to paint portraits with light. On a sunny day the cinematographer would first need to decide where to put the camera. The most common strategy was to face the sun, using it to produce a beautiful backlight effect, but some cinematographers preferred to use the sun as a cross-frontal key. Either way, the cinematographer would need to fill in the resulting shadows on the face. One option was to use arc lights to do this, although then a generator would be required if the exterior was on location. More commonly, the crew set up reflectors to bounce the sunlight back toward the performers. A textured reflector would soften the light; a smooth reflector would act like a mirror, maintaining the hardness of natural sunlight. A careless cinematographer might set up the reflectors on the ground, producing an unpleasant light from below. A more polished solution was to put the reflectors on stands, to mimic the height of studio lights. There is a certain irony about this development. Artificial lighting was originally designed to look like daylight; by the twenties, Hollywood cinematographers were manipulating daylight to look like artificial light.

Because of the increasing range of options, we might expect all cinematographers to have adopted differentiating strategies during the 1920s. However, another source of pressure encouraged greater homogenization: the cinema star phenomenon, which placed a high priority on glamour. Occasionally a cinematographer might glamorize a star by emphasizing an unusual feature, but it was more common to think of glamour as an ideal of perfection, with all unusual features smoothed away. As cinematographers grew better and better at making people look perfectly unblemished, the studios grew more and more insistent that cinematographers use their glamorous techniques on all major stars, men and women alike. When women are lit according to this ideal, the lighting simply reinforces the already established ideals of feminine beauty. By contrast, the problem of lighting a man could pull the cinematographer in two different directions. The ideals of male characterization would encourage him to employ a relatively harsh technique, but the ideals of glamorization would necessitate the use of a lighting treatment that was conventionally coded as "feminine": a flat key-light, a generous amount of fill, and a glowing backlight. This is particularly true for "Latin lover" types like Ramon Novarro and Rudolf Valentino. As scholars like Miriam Hansen and Gaylyn Studlar have argued, these men represented a distinctive ideal of masculinity.[37] Their ethnic identities separated them from more conventional male stars, and their faces and bodies were put on display in a manner that we more commonly associate with female stars. In some cases this could produce difficult problems for the cinematographer. In the Rudolf Valentino film *The Eagle* (Clarence Brown, 1925), the lighting fluctuates between movie-star glamour and tough-guy modeling. This suggests that cinematographer George Barnes was trying to follow two conventions at the same time. The action-packed narrative was calling for character lighting, but the star's matinee idol identity was demanding more "feminine" treatment.

It is clear that in the 1920s figure-lighting conventions were more complicated than the male/female binary would suggest. I suggest that they operated on three different levels, comprising a stratified hierarchy of increasingly complex lighting paradigms. At the first level, many cinematographers simply executed the basic technique of three-point lighting without trying to find variations that would differentiate between men from women. Put bluntly: these cinematographers lit everyone in the same way. Male or female, old or young, action star or glamour queen—everyone would receive the same arrangement of key, back, and fill. This may sound terribly uncreative, but we should remember that a cinematographer who had mastered the three-point lighting technique was already capable of fulfilling two very important functions: the illusion of roundness, and the production of a flattering image.

At the next level we can find cinematographers who embraced the portrait photographer's strategy of using visual style to differentiate the images of men and women. This was a more complicated strategy, but it did not require the use of unusual lighting schemes like core-lighting or double key-lighting. Instead, a cinematographer could simply make minor modifications to the basic three-point set-up. One option was to change the placement of the key-light. A frontal light would smooth out the features of a female subject, and a cross-frontal light would enhance the modeling for a man. Adding a little diffusion to a woman's key-light was another way to produce differentiation while remaining within the boundaries of the three-point style. By using this basic differentiation strategy, the cinematographer could take advantage of the roundness offered by three-point lighting, while adding the functional benefit of characterization.

Of course, this second-level kind of characterization was not particularly rich. Indeed, the binary logic rested on a disappointingly simple ideological opposition, softening the images of women and sharpening the images of men. At the third, most sophisticated, level of this categorization, cinematographers attempted to complicate this binary logic. A cinematographer who had thought carefully about the discourse of characterization would realize that Hollywood does not tell stories about men and women, but about seductive vamps and suffering mothers, about tough guys and Latin lovers. These are still thoroughly stereotypical representations, but they give the ambitious cinematographer another opportunity to prove his skill. Here the discourse of characterization intersects with the strategy of maximizing degrees of differentiation. A master cinematographer will recognize distinctions between different kinds of characters, and know how to represent those distinctions by crafting subtle variations on the basic norms. An aging mother should receive harder treatment than a vamp; a romantic male lead should receive softer treatment than a gangster.

My Best Girl provides an example of this more complex approach to characterization. During the first meeting between Maggie (Mary Pickford) and Joe (Charles "Buddy" Rogers), Rosher uses a more frontal placement for Pickford's key, while keying Rogers with a side-light (figures 2.16 and 2.17). This was a fairly routine handling of the male/female pairing, providing more modeling for the male performer. Rosher complicates this technique when he shows Joe in a scene with his father. Here, the relevant distinction is not male/female, but old/young. Rosher uses a hard light on the father, while shifting to a softer technique for the son (fig. 2.18 and 2.19). This is a simple shift, but it shows that Rosher refused to light Rogers the same way in every scene. For the advanced cinematographer, the lighting style changes to accommodate the shifting demands of characterization.

Clarence Brown's *The Goose Woman* (1925, shot by Milton Moore) is another example where lighting norms are relative. In the first half of the film, the title character (Louise Dresser) is an aging alcoholic who once had a career as a famous opera singer. For these scenes Moore lights Louise Dresser with harsh cross-lights. Rather than smooth out the wrinkles with a soft front light, the cinematographer emphasizes them in order to highlight the protagonist's lack of glamour. In the second half of the film, the onetime opera star gets the chance to reclaim her lost status. As Dresser is remade into a sophisticated woman, Moore switches to softer treatment. Here again the filmmakers modify the conventions to suit the particular needs of the story, altering the lighting for a single character within the story.

These are elaborations on the basic norms, rather than departures from them. The lighting in *My Best Girl* is still structured by the logic of character, with smooth lighting for the young, innocent Maggie, harder

Figure 2.16 A frontal key for Mary Pickford in *My Best Girl*.

Figure 2.17 In the same scene, Charles Rogers's key-light is coming from the side.

lighting for the wise old father, and variable lighting for the young-but-still-masculine Joe. Meanwhile, the hard lighting on Dresser in *The Goose Woman* functions as a character-specific variation on the established norm for lighting women. The lighting makes sense against a background of conventions.

⊚ ⊚ ⊚

All Hollywood cinematographers—even the most experimentally inclined—were using three-point lighting as a default convention by the 1920s. There is a temptation to say that this is proof of the power of norms, that three-point lighting was the classical norm, and that, one by one, the cinematographers of Hollywood acquiesced to the norm. In my view, this description is not quite correct. It implies that the norm acts as a kind of independent, external pressure, forcing cinematographers to adapt. Phrased in this way, there is too much emphasis on the power of

Figure 2.18 *My Best Girl*: Gentler modeling emphasizes Rogers's youth.

Figure 2.19 *My Best Girl*: A hard light from the side makes the father look old.

the norm to control a cinematographer's choices, and too little emphasis on the fact that it was filmmakers themselves who helped to create the norm. In the case of lighting, the filmmakers with the most influence were, in fact, cinematographers. As we have seen, cinematographers valued art and efficiency. To achieve efficiency, cinematographers needed to have a default convention. To some extent, any default convention would do; the point is that a default convention allowed them to light more quickly. Of all the possible default conventions, they chose three-point lighting because it was deemed the most capable of fulfilling two distinct aesthetic functions that they wanted to accomplish: namely, characterization and the illusion of roundness. When elite cinematographers like Rosher and Seitz decided to use three-point lighting as often as their peers, it does not mean that they had abandoned their personal artistic ambitions in the face of Hollywood homogenization. If anything, it proves that the artistic ambitions that they shared had raised the level of

Hollywood lighting as a whole. Now, all cinematographers were sharing a technique that could fulfill the ASC's institutionally endorsed ideals of art and efficiency. In short, cinematographers did not show their weakness by succumbing to a norm; they exercised their power by creating one.

The Drama of Light

3

I n his remarkable book *La lumière au cinéma* (Light in the Cinema),
Fabrice Revault d'Allonnes writes, "In classicism, one makes light with
a theatrical spirit, if not a theatrical essence. Expressive, rhetorical: dramatized, psychologized, metaphorical, and selective."[1] This argument
proposes a bold new way to look at Hollywood lighting. In the view of
some critics, classical filmmakers were so devoted to the ideal of invisibility that they produced a neutral, inexpressive style. For Revault d'Allonnes,
expressivity is one of the core values of the Hollywood cinema.

The Hollywood style is an intensely dramatic style. While several of the
most respected members of the American Society of Cinematographers
(ASC) were former portrait photographers, it was not common for cinematographers to have a background in the theater. Instead, cinematographers absorbed theatrical ideas in a mediated form, responding to the
ideas of theatrically trained directors like D. W. Griffith and Cecil B. De-
Mille. These directors drew on the lighting strategies of the theater as part
of a conscious effort to elevate the cultural reputation of the cinema. This
agenda overlapped neatly with the cultural agenda of the ASC. By shaping their lighting strategies according to a specifically dramatic conception of light, cinematographers made a forceful assertion about the expressive artistry of Hollywood cinematography.

A 1920 article in *Theatre Arts Magazine* offers a good introduction to
the theatrical approach to light.[2] In this article, theater director (and future filmmaker) Irving Pichel writes:

> Light, in the theatre, then: (1) illuminates the stage and actors; (2) states hour,
> season, and weather, through suggestion of the light effects in nature; (3)
> helps paint the scene (stage picture) by manipulation of light and shadow and

by heightening color values; (4) lends relief to the actors and to the plastic elements in the scene; and (5) helps act the play, by symbolizing its meanings and reinforcing its psychology.[3]

According to Pichel, one of the accomplishments of modern theater was the discovery of the multifunctional possibilities of light. For centuries the primary function of light had been simple illumination. Only recently had theater directors learned to use light for other purposes—to denote time and place, to enhance the beauty of the stage picture, to provide three-dimensional modeling for the actors and sets, and to enrich the significance of the play with symbolism and mood lighting.[4]

We can find similar ideas echoed in many other sources. In a 1929 book on stage lighting, Theodore Fuchs quotes the same passage, and then refers to it repeatedly throughout the book.[5] A 1930 book proposes a similar list, covering all five of Pichel's main functions, while adding the function of directing the audience's attention to the main point of interest.[6] A 1936 article offers an efficient list of four functions: visibility, locale, composition, and mood.[7] The fact that some of these texts were intended for amateur readers is an indication that these ideas were located comfortably in the mainstream of theatrical thought.

These authors suggest various strategies that allow the ambitious lighting designer to accomplish any or all of these functions. Two strategies stand out. First, a lighting designer might attempt to light the play in such a way that all of the light would appear to come from a source within the fictional world, such as a window or a table lamp. Some authors described this strategy as the creation of an "effect," as in the effect of sunlight or the effect of a flashlight. Second, a lighting designer could select a particular lighting atmosphere to suit the mood of the play: a farce might call for bright lighting, while a tragedy might merit a more somber style. In theory, the strategies of effect-lighting and expressive lighting are quite distinct; in practice, they can overlap in various ways.

In a section entitled "Effects," Fuchs writes, "Simulated effects utilize light as an integral part of the realistic scene, as a whole, that they help to create, and include principally a representation, or rather imitation, of such things as sunlight, moonlight, the sky, fog and mist, lightning flashes, fireplaces, camp fires, torches, lighted lamps, and so forth."[8] Fuchs goes on to detail various techniques. For instance, his description of the fireplace effect is remarkably thorough, explaining how to simulate logs, how to simulate smoke, and how to simulate a grate. Most narratives would not require this amount of detail, but a naturalist play might demand accuracy regardless of narrative concerns. Some theorists would link the concept of realism to an additional function, suggesting that a realistically detailed world produces the illusion of presence. According to this notion, the spectator looks into the fictional world, as if through an invisible fourth wall. Perhaps thinking along these lines, Fuchs himself suggests that the goal of the realist technique is "complete illusion."[9] On the other hand, a particularly spectacular effect might produce the opposite, shattering the illusion by drawing attention to the artfulness of the entire theatrical composition. In general, it is a good idea to think of an effect's functions—denoting time and place, enriching the world with realistic detail, producing the illusion of presence, adding beauty to the composition—as distinct goals, even if they can occasionally be combined in practice. It is clear that this strategy could be attractive to cinematographers, who were anxious to expand the functionality of light.

In addition to effect-lighting, the discourse and practice of the theater introduced another concept that would exert a strong influence on cinematographers: the idea that lighting style could be varied to suit the mood of each particular story. During the 1910s and 1920s, historians of the theater would have attributed this idea to Edward Gordon Craig (although Adolphe Appia was another important innovator in this regard). Craig directed only a handful of productions, but he was widely admired for his passionate arguments calling for a new style of theater. According to Craig, the theater director should strive to produce a unified effect. All the aspects of *mise-en-scène*—from acting to set design to lighting—should work together to enhance the expressive power of the play. Significantly, Craig scoffs at the naturalist theater. In one essay from 1911, he writes, "Actuality, accuracy of detail, is useless upon the stage."[10] Instead of imitating the lighting effects of the real world, the light of the theater should express the deeper emotional truths of the play. In other words, Craig argues that his expressive lighting strategy is not compatible with the detail-oriented techniques of effect-lighting. The two strategies serve different goals.

The conventions of expressive lighting were not very well developed. Writing in 1929, Fuchs laments that most expressive lighting follows a fairly predictable pattern. "The rule of thumb seems to have been: bright light in full blast for comedy and farce, a dim light for deep tragedy, and all the proportional gradations of lighting for the intervening range of emotions."[11] This is admittedly a rather simplistic system, but we should not underestimate its power. A designer who had mastered this simple guiding precept could make some very bold assertions about the artistry of lighting. Following Craig, the lighting designer could claim to have realized the aesthetic ideals of harmony, insight, and expression.

Craig preferred expressive lighting to the effect-lighting of realist theater—in particular, the detail-obsessed lighting of naturalism. In the work of some directors, however, these apparently antithetical approaches began to converge. David Belasco is a case in point. Belasco was a highly successful director who had taken naturalism to notorious extremes when he recreated the interior of Child's Restaurant in New York for his 1912 production of *The Governor's Lady*. Throughout his career, Belasco challenged his lighting designer, Louis Hartmann, to develop unusual lighting effects, such as the effect of a sunset gradually changing colors over the course of a scene. Although Belasco's plays were financially successfully, he was often the target of criticism—in particular, criticism from writers who were influenced by the avant-garde theories of Craig. For instance, Sheldon Cheney, a Craig disciple, wrote a sharp critique of Belasco for a 1914 issue of *The Theatre*. Like many critics of the period, Cheney was skeptical of "photographic" realism—a naturalistic realism that offered a quasi-scientific imitation of superficial details. Accurate effect-lighting could be interesting in a scientific way, he felt, but it made no contribution to the development of the theatrical art.[12]

This was not Cheney's only complaint against the Belasco style. After comparing his work to a science experiment, Cheney made an even more demeaning comparison: to vaudeville. Just as a vaudeville show might use a simplified narrative "as an excuse for introducing singing, and anecdote-telling and episodic happenings without regard to cumulation of dramatic interest," Belasco disregarded the needs of the story in order to draw attention to a stunning piece of set design or to a spectacular new lighting effect.[13] This brings us back to the point that effect-lighting could accomplish four distinct functions, such as adding realistic detail and producing the illusion of presence. We normally think of these functions as being related. By giving more details to the fictional world, an artist can increase the illusion of presence. Here the two functions are at odds, because realism itself is being put on display. The spectator is being asked to compare the light of the theater to the light of the world. This act of comparison forces the spectator to think about the effect as a representation, and not as part of a unified whole. The end result, according to Cheney, is a weakening of the play's effectiveness: whereas Craig had called for a theatre of synthesis, with light contributing to a greater harmony, Belasco offers a theatre of strikingly realistic but thoroughly disconnected effects.

Cheney's article reinforces the opposition between Craig's expressive lighting and Belasco's effect-lighting. In his 1919 book *The Theatre Through Its Stage Door*, Belasco responds to his critics. In the process, he argues that his recent work had transcended that very same opposition:

> My method of presenting plays was never without its strong advocates. The latter saw more clearly than my adverse critics. They divined that the careful attention I gave to the extraneous details of my productions was only for the purpose of intensifying and interpreting the mood of the play and of the characters, and that I was trying by legitimate artistic means to stir the emotions of my audiences.[14]

We see that Belasco agrees with Craig's basic principle: in order to maximize the play's expressive power, the lighting should be varied to suit the changing moods of the play. However, Belasco insists that he can accomplish this goal without abandoning his abiding commitment to naturalism. His solution is to find plausible "pretexts" to motivate each effect. For instance, if the emotional quality of the scene calls for an orange hue, then Belasco can set the scene at sunset. If the scene calls for darkness, then Belasco can have an actor flip off a light switch. These pretexts allow Belasco to employ the techniques of effect-lighting, while simultaneously matching the mood of the story.

In response to accusations that he is little more than a master of eye-catching spectacle, Belasco argues that his pretexts work to conceal the artistry of his accomplishments. A well-motivated lighting effect does not distract the audience's attention from the play. Instead, it sharpens the audience's focus by maintaining the coherence of the fictional world. Belasco writes:

> There are also thousands of chances for delicate strokes of illumination in a well-managed modern play which neither audience nor critic is likely to notice, yet which work unconsciously upon the feeling and imagination.
>
> To select the right opportunities for their use, to know how to contrive them, and at the same time how to conceal them, is what makes the profession of stage director so difficult. Not only should he have a comprehensive knowledge of all the arts, he must understand psychology and the physical sciences besides. In the intricate process of producing a play he must be the translator of its moods, and supply the medium by which they are transmitted to audiences.[15]

Belasco is probably being a little disingenuous here, since he was a brilliant showman who knew how to use spectacle to draw in a crowd. Still, he is articulating an important idea. To our original list of the four functions of effect-lighting, Belasco has added a fifth: with careful handling, effect-lighting can function as expressive lighting, enriching the emotional impact of the play. In so doing, Belasco has broken down the opposition between realistic lighting and expressive lighting.

Before we give Belasco too much credit, we should consider one additional fact: he was not the only person at the time who used effect-lighting for expressive purposes. In fact, the idea was probably quite commonplace in one of the theater's most popular genres, the crime-story melodrama. The 1912 melodrama *Within the Law*, by Bayard Veiller, offers a particularly dense example. One scene represents a tense physical altercation. The stage directions call for a series of light effects. First, a character flips off the lights. Then another character enters and stands in the light of a doorway. The door closes, and a struggle ensues in the darkness. A flashlight is turned on, then off again. Finally, a character turns on a lamp, illuminating the faces of two characters, while keeping the rest in shadow.[16] It is impossible to know exactly what this scene looked like onstage, but it is clear that Veiller wants this crime scene to take place in an atmosphere of menacing darkness, punctuated with strategically selected light effects. Similarly, in the 1922 comic melodrama *The Cat and the Canary*, the first act ends with a suspenseful situation, described in the following stage directions: "Annabelle, sensing danger, reaches up and turns off the light, leaving herself in the dark with the maniac. She rushes to the door through the moonlight from the windows, opens it and dashes out, slamming the door."[17] Rather than simply calling for darkness in the stage directions, the writer has given the director a set of pretexts, motivating the rapid shifts from light to shadow. The lighting adds realistic detail while establishing an expressive mood.

The discourse and practice of theater lighting had much to offer an institution like the ASC, which was committed to expanding the artistic resources available to the cinematographer. However, it is not possible to draw a direct line between a theatrical showman like David Belasco and a Hollywood cinematographer like Charles Rosher. While several important cinematographers had worked in the field of portrait photography, the theater was not an important training ground for the typical ASC member. Instead, film directors acted as a mediating influence. During the 1910s and 1920s, some of the most important directors in Hollywood were men with experience in the theater, such as D .W. Griffith. Cecil B. DeMille and his brother William C. de Mille had experience with David Belasco himself. While in France, Maurice Tourneur had worked in the theater of the naturalist director André Antoine.[18] In Germany, Ernst Lubitsch acted in the company of Max Reinhardt, a brilliantly eclectic director who mixed elements of realism, symbolism, and expressionism.[19] Some directors had less illustrious résumés. Roland West wrote melodramatic plays about criminals and the police.[20] Tod Browning worked on the fringes of the legitimate theater—in the circus, in a magic act, in vaudeville.[21] All of these directors would bring theatrical ideas to the cinema—including ideas about the dramatic handling of light.

Because of his close links to David Belasco, Cecil B. DeMille offers a good example of a director who incorporated theatrical concepts into the art of film lighting. DeMille hired the theatrically trained Wilfred Buckland as his art director, and Buckland worked with cinematographer Alvin Wyckoff to develop a wide range of lighting effects for such films as *The Cheat, The Golden Chance*, and *The Heart of Nora Flynn* (all 1915). The opening sequence of *The Cheat* provides a famous example. This sequence introduces us to the film's villain, Tori (played by Sessue Hayakawa).[22] Tori brands his possessions, and he is heating his branding iron in a brazier. Buckland and Wyckoff stage an elaborate lighting effect, suggesting that the fire in the brazier is providing the primary illumination on Tori's face (see fig. 3.1). The detailing is impressive: when Tori blows on the fire, the brightness of the light fluctuates, to suggest the momentary flaring of the flames. We can imagine that Belasco himself would have been proud of such details.

Figure 3.1 Effect-lighting during an introductory shot from *The Cheat* (1915).

As we have seen, Belasco's style fulfilled multiple functions, but those functions could pull in different directions, sometimes aiding the story, sometimes distracting from it. A similar kind of tension shapes the lighting effects in *The Cheat*. On the one hand, the effects perform storytelling functions; on the other hand, they perform functions that are directly at odds with the needs of efficient narration. As Lea Jacobs has argued, the early DeMille style may have influenced the classical style, but we must remember that it is not identical to it.[23] Specifying the functions of DeMille's effects can sharpen our understanding of the changes to come. First, we can group some functions under the heading of "storytelling." While the example from *The Cheat* is an introductory shot, prior to the beginning of the causal chain, it does give us information that will help us understand the narrative to come. With the aid of some tinting and toning, the dark background suggests that the time is night, and the flickering of the light provides denotative information about the flames. Tori will later be associated with both motifs: the night and the flames. Another basic storytelling function is the control of the spectator's attention. In figure 3.1, the bright light calls attention to Tori's face and to the branding equipment, both of which will play important roles in the story. Meanwhile, the darkness of the background prevents the appearance of distracting details.

These are routine narrational functions, but lighting can also intensify the drama in an expressive way, enhancing the mood of the story. In a 1916 interview, DeMille argues this point explicitly:

I have found that emphasizing or softening certain dramatic points in the motion picture can be realized by the discriminating use of light effects, in just the same way that the dramatic climax in a play can be helped or impaired by the music accompanying it, and working on this principle I came to feel that the theme of the picture should be carried in its photography.

In our production of "The Cheat," one of the principal characters is a Japanese. In photographing this I endeavored to carry out the Japanese school of art by making my backgrounds sinister and using abrupt, bold light effects. In fact, the lighting of this picture definitely suggests the "clang" and smash of Japanese music. . . .

As a general thing, light effects are out of place in comedy. There you will notice our lighting is clear and brilliant, corresponding to the faster light comedy theme in music, except in the melodramatic scenes, where we carry our audience into thrills, not only by the action of the artists, but by a change in the mood of our photography.[24]

Like Appia, DeMille compares light to music. Like Belasco, DeMille sees no opposition between realistic lighting and expressive lighting. DeMille's effects are realistically motivated, but that does not prevent them from accomplishing an expressive function. Admittedly, DeMille's explanation in this passage rests on some absurd assumptions. What do sinister backgrounds and abrupt effects have to do with Japanese art? However, the absurdity of the explanation does help to lay bare the underlying logic. For DeMille, expression works via associations. More specifically, expression works via *differentiated* associations. A lighting technique stands out because it is different from other lighting techniques, and it produces associations that are equally differentiated. It is significant that Belasco also supports his theory of expressive effect-lighting with an example from a play set in Japan. Belasco and DeMille are trying to prove that different stories call for different kinds of light. For both men, a story involving the Japanese seems to be the ultimate example of a different kind of story; they use racial difference as a convenient shorthand for differences in narrational strategies. The theory of expression seems like a purely aesthetic theory, but ultimately it is a theory of differentiation. As such, it rests on culturally specific notions of difference.[25]

The link between DeMille's abrupt lighting effects and his story about a Japanese villain may be a tenuous link, but it shows that DeMille is committed to providing a connection between style and story. However, as Jacobs insists, we must acknowledge that storytelling is not the only function that light effects can perform. They also serve nonnarrative functions, like pictorial display. A properly tinted and toned 35mm version of *The Cheat* is a beautiful thing indeed, worth seeing regardless of the story. When he told his oft-repeated story of the invention of "Rembrandt lighting," DeMille himself acknowledged this complex combination of narrative and nonnarrative appeals:

I will show you the birth of artificial lighting. When we first went to California everything was sunlight. No artificial light was employed. Having come from the stage I wanted to get an effect, so I borrowed a spotlight from an old theater in Los Angeles when I was taking a photograph of a spy in *The Warrens in Virginia*. The spy was coming through a curtain and I lighted half of his face only, just a smash of light from one side, the other side being dark. . . .

When I sent the picture on to the sales department I received the most amazing telegram from the head of this department saying, "Have you gone mad? Do you expect us to be able to sell a picture for full price when you show only half of the man?" . . . I sent a telegram to New York saying, "If you fellows are so dumb that you don't know Rembrandt lighting when you see it, don't blame me." The sales department said, "Rembrandt lighting! What a sales argument!" They took the picture out and charged the exhibitor twice as much for it because it had Rembrandt lighting. That is the history of artificial light in motion pictures today.[26]

Ever the showman, DeMille exaggerates his role in the birth of cinematic lighting. He was not the first to use artificial light, and he was not the first to introduce ideas from the theater. Still, this is a valuable passage because it reveals the way lighting can offer multiple appeals. In the first paragraph, DeMille argues that the lighting performs a dramatic function, setting the right mood for a scene of spying. In the second paragraph, DeMille jokingly explains how lighting can be turned into an appeal that can be sold for its cultural value. With the right sales pitch, "Rembrandt lighting" can function as an obvious marker of pictorial quality, an attraction in its own right.

Some early lighting effects take this pictorial strategy a step further, mirroring even more specific painterly techniques. For instance, Maurice Tourneur often groups some characters around a table with a hidden light-source in the middle. Figure 3.2 offers an example from *The Last of the Mohicans* (1920, co-directed by Clarence Brown). This arrangement openly borrows a compositional strategy that was popular in seventeenth- and eighteenth-century painting, appearing in the works of such painters as Georges de la Tour and Joseph Wright of Derby.

Jacobs has argued that this spectacular quality is one of things that differentiates the lighting effects of the teens from the more invisible lighting techniques of the classical Hollywood cinema. She points out that DeMille's lighting effects are often *over*-motivated. The characters go

Figure 3.2 A lamp effect imitates a technique that was common in seventeenth- and eighteenth-century painting.

FROM *THE LAST OF THE MOHICANS* (1920).

through elaborate motions to turn lamps on and off, thereby providing an overly obvious justification for the spectacular effects. I would add that this extra justification does not render the techniques extra-invisible. In fact, the opposite is the case. When Tori blows on the fire, and the light brightens and dims to indicate the flaring of the flames, it is the careful motivation itself that provides the primary point of interest. A man blowing on a flame is not very interesting. Neither is the momentary brightening and dimming of a light. But the perfect coordination of the two is interesting, precisely because of the conspicuous accuracy of the effect. This is realism on display.

No matter how much he protested that his lighting effects helped his stories, Belasco was always subject to the criticism that his overly conspicuous realism was a distraction, drawing attention away from the story. DeMille could be subject to the same charge. However, Jacobs rightly insists that is would be somewhat ahistorical to make such a charge. DeMille was not combining narrative and spectacle because he was defying the dictates of classicism, for the simple reason that those dictates were still in the process of formation. At this point in history (1915), filmmakers like DeMille were admired precisely for their ability to combine narrative and spectacle in an emotionally engaging way.

By the late teens, DeMille had moderated his aggressive lighting style, though he continued to use effects at strategic moments. Most of the lighting in *The Affairs of Anatol* is fairly flat, but the climactic scene uses a striking low-placed key-light effect to heighten the atmosphere of evil as the protagonist visits the residence of a woman named Satan Synne. Indeed, DeMille continued to use dramatically appropriate effects throughout his career.

When the ASC was formed, its members could look to the films of DeMille, Griffith, and Tourneur as models of modern film lighting. These directors had worked hard to elevate the cultural standing of the cinema, making it even more attractive to a middle-class audience.[27] As discussed in chapter 2, the ASC had cultural ambitions of its own, hoping to create a new public identity for the cinematographer as a highly skilled professional artist. Adopting the rhetoric of culturally ambitious directors was an obvious way to advance that agenda.

Many of the effects introduced by Griffith, DeMille, and other theatrically trained directors soon became commonplace, changing from one-off gimmicks to widely accepted conventions. For instance, D. W. Griffith and Billy Bitzer had experimented with lighting effects in their early shorts. Two well-known examples are the fireplace effect in *A Drunkard's*

Reformation (1909) and the effect of lighting streaming through a window in *Pippa Passes* (1909). By the 1920s, the fireplace effect had become fairly common, appearing in films like *The Toll Gate* (1920) and *Flesh and the Devil* (1926). The window pattern was even more familiar. Figure 3.3 gives a typical example, from *Our Hospitality* (John Blysone and Buster Keaton, 1923, photographed by Gordon Jennings and Elgin Lessley). Here the technique accomplishes three functions: imitating the appearance of sunlight, adding a dash of pictorial beauty, and emphasizing an important element of the scene, the weapons. A cinematographer could vary the window pattern in several ways. Dimming the general illumination would create the impression of moonlight glowing in a darkened room, as in the prison scene from *The Mark of Zorro* (Fred Niblo, 1920, photographed by William McGann and Harry Thorpe). Occasionally the light would take on symbolic associations: in *The Eagle* (Clarence Brown, 1925), George Barnes turns the window pattern into a symbol for the divine. In *The Wind* (Victor Sjöström, 1928), John Arnold reverses those connotations, using window patterns to remind viewers of the malevolent force represented by the howling winds outside.

Some other common effects involved the street lamp, the table lamp, the swinging lamp, the handheld candle, and lightning. In *The Bat* (Roland West, 1926), cinematographer Arthur Edeson uses a narrowly focused spotlight to imitate the effect of a woman lit by a candle. In figure 3.4, from *Old San Francisco* (Alan Crosland, 1927), cinematographer Hal Mohr places a strong source within the lamp held by the man in the background. He supplements this light with an even more powerful source, providing the bright "kicker" on the man in the foreground. Later in the same scene from *Old San Francisco*, Mohr uses another effect: a cast shadow, projected on the wall behind Dolores Costello as she watches two

Figure 3.3 A window pattern in *Our Hospitality* (1923).

Figure 3.4 A handheld lamp in *Old San Francisco* (1927).

Figure 3.5 *Old San Francisco*: The same lamp motivates the cast shadow during a fight.

Figure 3.6 A shadow cast by an offscreen character in *The Three Musketeers* (1921).

men fight (fig. 3.5). In this example Mohr is careful to motivate the effect: we are supposed to assume that the lamp from figure 3.4 is casting this shadow. However, cast shadow effects did not always receive such careful motivation. In *The Three Musketeers* (Fred Niblo, 1921), for instance, cinematographer Arthur Edeson uses an unmotivated cast shadow to suggest the nefarious presence of Cardinal Richelieu, on the right (fig. 3.6). Cinematographers generally preferred hard lights for the cast shadow effect, since it was important to give the shadows crisp definition. (Note that "hard lighting" is not the same thing as high-contrast, or low-key, lighting. A light is hard when it produces a shadow with a sharp outline. The cinematographer can keep the shadow dark by using minimal fill, as in 3.5, thereby producing a low-key, hard effect, or he can brighten the shadow with extra fill light, as in 3.6, thereby producing a high-key effect with shadows that are still equally hard.)

Backlighting may have started out as an effect, imitating the appearance of sunlight in a woman's hair. Although this technique soon became a routine figure-lighting convention, it could occasionally function as an effect, with the proper motivation. For instance, in a suspenseful scene from *The Wind*, the backlighting is part of an elaborate swinging-lamp effect.

Another popular effect derived from the silhouette. In figure 3.7, from *The Love Light* (Frances Marion, 1921), cinematographers Charles Rosher and Henry Cronjager imitate the look of a moonlit exterior by combining a narrow aperture with an absence of fill light, while using a filter to darken the sky. The two lovers merge into the nocturnal landscape, in an artful composition that brings together various romantic motifs: the tree, the sky, and the ocean. Pictorialist photography probably inspired this silhouette technique. Alfred Stieglitz's famous photography journal *Camera Work* had published several images with similar compositional strategies, such as J. Craig Annan's 1904 photograph "The Dark Mountains," or Annie W. Brigman's 1912 photograph "Dawn."[28] This magazine championed the idea that photography was a creative art, rather than a technique of mechanical reproduction.[29] This helps explain why the silhouette was popular with the ambitious cinematographers of the ASC. However, the silhouette has limited value for storytelling, since it keeps the actors in shadow, often in long shot. Another disadvantage is the fact that a silhouette, unlike chiaroscuro, fails to model the object, thereby detracting from the illusion of roundness.[30] Perhaps because of these limitations, the silhouette was saved for very specific situations, like exotic locations and romantic embraces.

At the most general level, we can describe the light as an effect whenever the lighting scheme imitates the appearance of a particular time and place. Alleys and underground passageways are generally considered dark places. In such locations, a dark lighting scheme would produce the appropriate effect, whether the cinematographer chose to show a particular source or not. For instance, in *The Penalty* (Wallace Worsley, 1920),

Figure 3.7 A silhouette effect in *The Love Light* (1921).

one scene shows the protagonist investigating the hidden chambers of Blizzard, a master criminal played by Lon Chaney. Cinematographer Don Short uses a high-contrast setup, creating the dark shadows that we normally associate with the hidden chambers of criminal masterminds.

Drawing on the theory that the Hollywood style was a "classical" style, we might assume that Hollywood cinematographers favored motivated effects because motivated effects were more likely to be "invisible." To some extent, this is a useful assumption. Lighting effects could perform routine narrational functions in an unobtrusive way. Because narrative involves events occurring in space and time, any storyteller must learn how to communicate information about space and time efficiently. A flickering candle effect can communicate such basic information instantly. As James Wong Howe once said, the cinematographer "has to indicate the morning hours without bringing in a crowing rooster. Have the right lighting for evening without showing a man walking home with his lunch box."[31]

However, we should balance this "classical" conception of effect-lighting with an awareness of two other factors. First, not all Hollywood cinematographers were fully committed to the ethic of invisibility during the silent period. The ASC encouraged cinematographers to think of themselves as artists, thereby endorsing a certain amount of stylistic spectacle. Second, one could make a plausible case that strong motivation within the diegetic, or storytelling, world made a lighting effect *more* visible, not less. Indeed, a British photographer once suggested as much in a 1912 article: "For lamplight effects a standard or table lamp will be necessary, and in every case the lamp must be shown in the picture, otherwise the 'motif' is absent and the lighting borders on the ordinary, no matter how cleverly managed."[32] The lighting effect does not disappear when it appears realistic; rather, the lighting effect encourages spectators to admire the conspicuous accuracy of the effect. Elaborate effects put realism on display. The examples from *Old San Francisco* and *The Wind* show that spectacle was still an important function for effect-lighting, over a decade after *The Cheat*. We should also remember that filmmakers used tinting, toning, and other lab techniques to make effects look even more spectacular. William Daniels tells a story about a lost scene from *Greed* (Erich von Stroheim, 1924): lab workers spent months coloring a candle flame with a one-hair brush—a laborious task they had to do separately for every print and frame![33]

The conventions of effect-lighting were not legalistic mandates that applied to every situation. Instead, they were simple guides that developed in an *ad hoc* manner. One creative cinematographer might invent a new moonlight effect, and another cinematographer would copy the trick. A third cinematographer might add a new variation, and a fourth would pass the technique onto a fifth. Eventually, via a daisy chain of informal influence, the precept would pass on to every cinematographer in Hollywood. The ASC existed, in part, to facilitate this very process. By convening meetings and social events, it brought the elite cinematographers of Hollywood together, giving them the opportunity to exchange their ideas. To be sure, some cinematographers may have preferred to hold on to their secrets, but the general agenda of the ASC was to make the tools of artistry available to all of its members. In other words, the fact that these techniques became conventions should not lead us to say that the forces of factory production had defeated the forces of artistic innovation. Quite the contrary: the fact that these techniques became conventions indicates that the ASC was successfully accomplishing its mission to elevate the artistic standards of Hollywood cinematography.

As we have seen, theatrical directors like David Belasco believed that the conventions of effect-lighting could be combined with the conventions for lighting a story according to the mood of its genre. It is easy to see why the ASC might have been attracted to this idea. The ASC wanted to create a public identity for the cinematographer as a highly skilled artist. The "right-mood-for-the-story" theory could help project this new identity in two ways. First, it would posit the cinematographer as a skilled worker with specialized abilities, such as the ability to interpret stories correctly and the ability to produce a wide range of lighting styles. Second, it would define the cinematographer as an artist who put those skills to work in the interests of a recognizable aesthetic goal—the telling of emotionally engaging stories.

Cinematographers soon developed an elaborate set of genre/scene conventions, establishing different moods for different kinds of genres and different kinds of scenes. The genre/scene conventions have a complicated relationship to the normative ideals of the "classical style." Critics have long acknowledged that the Hollywood style contains examples of expressive lighting, in such genres as the horror film and the film noir. However, these examples are generally seen as exceptions to the dominant norms of the invisible style. My research will propose a major revision to this account: expressive lighting was in fact *central* to the discourse and practice of the ASC. On the level of discourse, the ASC embraced the "right-mood-for-the-story" theory as a perfect tool for advancing its aesthetic agenda. On the level of practice, the elite cinematographers of the ASC developed and shared an elaborate set of expressive genre/scene conventions—conventions that can be seen in several major Hollywood films.

For instance, in 1924 *American Cinematographer* published a major article by Bert Glennon, cinematographer for DeMille's latest spectacular, *The Ten Commandments*. In this article Glennon argues that the cinematographer should never be content with the mechanical task of "getting an exposure on the film." Instead, the cinematographer should work with the director to determine the specific idea for each scene. Then he should modify the lighting to convey that idea. Glennon writes:

> An idea, we may say, is a thought or a manifestation of mind, and the camera is one of the means of its expression. . . . The first illustration is that of the opening episode of "The Ten Commandments." The idea there was—slavery—torture—broken spirit—depression and tyranny. Did that thought not reach the camera? And was it not enhanced by atmospheric photographic lighting?[34]

Glennon is adopting an "expression" theory of art. The work of art starts out as an idea in the mind of the artist; the artist's task is to communicate this idea to the audience. Glennon describes these ideas in emotional terms. The vehicle for those emotions is lighting: specifically, atmospheric lighting. Later in the essay, Glennon describes the process as "painting with arc lights."[35]

Glennon rejects the idea that this principle could be turned into a set of shorthand rules. However, if we look at some examples, it becomes clear that the cinematographers of the silent period were beginning to develop some widely shared conventions of expressive lighting. As in the theater, expressive lighting was particularly important for the genre of the melodrama. While scholars now refer to Douglas Sirk's family dramas of the fifties as melodramas, the term had a very different meaning in the teens and twenties. According to Steve Neale, "The mark of these films is not pathos, romance, and domesticity but action, adventure, and thrills; not 'feminine' genres and the woman's film but war films, adventure films, horror films, and thrillers, genres traditionally thought of as, if anything, 'male.'"[36] Following this account, the term "melodrama" will here refer to crime films, horror films, and other genres with intense scenes of action.

The high-contrast image with dark shadows is the most common convention of the melodramatic style. Although it is common to refer to shadowy cinematography as "expressionist," scholars like Barry Salt have shown that this convention became established well before Hollywood cinematographers borrowed any techniques from German Expressionism.[37] We have already seen (fig. 3.1) the theatrically trained DeMille using bold contrasts for melodramatic situations in *The Cheat*. Over the next few years, his studio, Famous Players–Lasky, continued to employ this style for melodramatic scenes—that is, for scenes of action, scenes of crime, and scenes of horror. In the 1917 movie *The Secret Game* (directed by Cecil's brother William C. de Mille), cinematographer Charles Rosher uses strong contrasts to photograph a suspenseful scene of spying. In a murder scene from *Stella Maris* (Marshall Neilan, 1918), Walter Stradling highlights Mary Pickford's eyes while casting shadows onto the rest of her face (fig. 3.8). In the John Barrymore version of *Dr. Jekyll and Mr. Hyde* (John Robertson, 1920), Roy Overbaugh typically uses stronger contrasts for the horrific Hyde scenes than he does for the less melodramatic Jekyll scenes. While this style was originally associated with Famous Players–Lasky, cinematographers at a variety of studios began using strong contrasts for melodramatic situations. Indeed, I have already mentioned several examples, such as the prison scene from *The Mark of Zorro* and the fight scene from *Old San Francisco* (fig. 3.5). The fact that these techniques were becoming conventions is demonstrated by the fact that *The Penalty* and *The Bat* both use shadows to represent the same basic situation: a woman discovering a hidden chamber while investigating a crime.

The conventions of melodrama often overlap with the conventions of effect-lighting. In a melodrama, people are constantly standing next to lamps, or blowing out candles, or walking past moonlit windows. These

Figure 3.8 Dark shadows for a melodramatic scene in *Stella Maris* (1918).

staging decisions give cinematographers the opportunity to motivate their generically appropriate shadows as realistically motivated lighting effects. The low-placed key-light is another melodramatic technique with roots as an effect. The great film noir cinematographer John Alton roots the origin of the technique in the American cinema of the teens, rather than in the German cinema of the twenties:

> Years ago, when in pictures we showed Jimmy Valentine cracking a safe, he usually carried the typical flashlight in one hand, while with the other he worked on the safe combination. In some scenes, the flashlight was placed beside him on the floor. In either case, the light source was established as a low one. To create an authentic effect, the cameraman lit the character from a low light which illuminated the face from an unusual angle. It distorted the countenance, threw shadows seldom seen in everyday life across the face. This light, which exaggerates features, became so popular that even in our films of today, when we want to call the attention of the audience to a criminal character, we use this type of illumination.[38]

According to Alton, the technique started as an "effect-lighting" convention, imitating the look of a flashlight. However, it soon became a genre/ scene convention, appearing in crime films whether the criminal had a flashlight or not. In Maurice Tourneur's 1915 version of the story, Valentine himself is not lit from below, but Tourneur introduces one of his criminal colleagues in a shot that looks like Alton himself could have shot it (fig. 3.9). The techniques worked for different kinds of criminals: in *The Toll Gate* (Lambert Hillyer, 1920), Joseph August uses a low-placed key to introduce William S. Hart's good-bad protagonist; in Tod Brown-

Figure 3.9 A criminal lit from below in *Alias Jimmy Valentine* (1915).

ing's *White Tiger* (1923, shot by William Fildew), Wallace Beery's thoroughly bad antagonist is introduced via a similar technique.

Another effect-lighting technique that cinematographers associated with the melodramatic style was the cast shadow. Because of images like figure 3.10, from F. W. Murnau's *Nosferatu* (1922, shot by Gunther Krampf and Fritz Arno Wagner), historians are fond of citing the influence of German Expressionism whenever a Hollywood film uses a cast shadow, but this technique was probably an established melodramatic convention before Hollywood cinematographers had seen Murnau's horror masterpiece, which was not released in the United States until 1929. Indeed, one image (fig. 3.11) from Harold Lloyd's 1920 comedy short *Haunted Spooks* (Alfred Goulding and Hal Roach) looks remarkably like the image from *Nosferatu*, though it predates Murnau's film by two years. Arthur Edeson was particularly fond of this convention, using it to represent a villain in *The Three Musketeers*, to introduce a murder scene in *Robin Hood*, and to suggest menace throughout the comic melodrama of *The Bat*.

Perhaps because of the cast shadow convention, the melodramatic style typically favors hard lighting. Even close-ups employ hard light. The plot of Mary Pickford's *Little Annie Rooney* (William Beaudine, 1925) mixes sentiment and comedy, but cinematographers Charles Rosher and Hal Mohr shift to a melodramatic style to film a murder scene. A character turns off the lights, and we see a series of lighting effects, with strong contrasts and cast shadows. A close shot of the murdered father uses remarkably hard lighting, sharply etching the lines of character in his face (fig. 3.12).

While *The Cabinet of Dr. Caligari* (Robert Wiene, released in the United States in 1921) may have reinforced some of these melodramatic lighting

Figure 3.10 A cast shadow in
F. W. Murnau's *Nosferatu* (1922).

Figure 3.11 A similar shadow in
a Harold Lloyd film from 1920
(*Haunted Spooks*).

Figure 3.12 Hard lighting for
a scene in *Little Annie Rooney*
(1925) in which the protagonist's
father dies.

conventions, we should remember that the film's lasting legacy is found
in its unusual art direction, rather than its lighting. Furthermore, the influence of German Expressionism is questionable when we consider that
there were several other more proximate sources for Hollywood's melodramatic lighting conventions. Clearly, the melodramatic theater was a

likely source—especially for films like *The Bat*, based on a play from 1920.[39] To this we can add much earlier examples from the graphic arts. In his autobiography Karl Brown cites Gustave Doré as an influence.[40] Doré's engravings frequently use carefully detailed lighting effects to represent criminal milieus: see, for instance, *The Bull's-Eye*, illustrating a policeman with a flashlight, or *Opium Smoking*, illustrating an opium den lit by a single candle.[41] One popular crime magazine, *The National Police Gazette*, often used shadowy imagery to represent murders and robberies: one 1893 illustration even shows an intruder's face lit by a gun blast from below.[42] Books with horrific subject matter could include similar illustrations, as in the shadowy illustrations typically included in editions of books by Edgar Allan Poe.[43]

I am not suggesting that cinematographers regularly perused the works of Poe, or collected back issues of *The National Police Gazette*. My point is simply that these generic conventions were already widespread in popular culture before the twenties—and well established in Hollywood before the U.S. release of many German classics in the latter part of that decade. Cinematographers did not need to turn to German Expressionism to learn to associate shadows with crime and horror; they could simply turn to long-standing traditions in domestic culture.

I do not mean to imply, however, that German cinema had no influence on the Hollywood style. In the area of camerawork, the influence was enormous. Murnau's *The Last Laugh* was released in early 1925, and E. A. Dupont's *Variety* came out in the summer of 1926. Karl Freund had photographed both films, and his brilliant camera moves received rave reviews in the American press. Several German directors came to work in Hollywood, as did Freund himself. Meanwhile, ambitious American directors began experimenting with "German" camera angles. The influence on camerawork is undeniable, but we should remember that institutions like the ASC worked as mediating factors. Initially, American cinematographers were not happy with the success of the Germans. Several articles in *AC* indicate that they were jealous of the attention Freund was receiving. Cinematographers worried that producers were giving coveted directing spots to German filmmakers. In one of his editorials, Foster Goss attacked a *Film Daily* article that had praised the use of camera angles in German films: "Many of the angles mentioned in the article in question, according to Gilbert Warrenton, A.S.C., would appear palpably crude if they were incorporated in American productions, for the simple reason that they were discarded as obsolete in the Middle Ages of cinematography."[44]

Anyone who has seen *Variety* or *The Last Laugh* knows that Goss and Warrenton were doing the German cinema an injustice. Warrenton was soon given the opportunity to atone for this insult, as Universal assigned him to shoot the film version of a classic theatrical melodrama, *The Cat and the Canary*, for German director Paul Leni. The 1927 film is packed with the bravura camera moves that had suddenly become a Hollywood vogue. Leni was probably responsible for these moves, since directors often had more control over camera angles. However, we should not necessarily assume that Leni was responsible for the moody lighting, even if his 1924 *Waxworks* is a good example of an Expressionist film that emphasized lighting as much as art direction. *The Cat and the Canary* was adapted from a stage play, and audiences would probably have recognized in its shadowy style the look of a theatrical melodrama. Indeed, Warrenton himself was one of the rare cinematographers with a background in theater, having come from a theatrical family of three generations.[45] Any account of German influence on Hollywood lighting must acknowledge that its influence merely reinforced techniques that cinematographers had already borrowed from the theater and other proximate cultural sources.

The melodramatic conventions were not the only genre/scene conventions used in Hollywood during the silent period. For instance, the romance can serve as the melodrama's polar opposite, featuring soft lights, gentle contrasts, and heavy layers of lens diffusion. In figure 3.13, from *The Mysterious Lady* (Fred Niblo, 1928), we see how cinematographer William H. Daniels uses lens diffusion to give the highlights a gentle glow, lamp diffusion to soften the shadows on Garbo's face and neck, and ample fill light to reduce the contrast. If we compare this image to figure 3.12, we see that the melodrama also intensifies the gendered lighting techniques of figure-lighting. This is not surprising, given the fact that the melodrama and the romance were both gendered genres: the melodrama was generally seen as a male genre, while the romance was often seen as female.

The comedy and the drama also had generic lighting conventions. Comedies generally employ bright lighting, even when the situation might call for a darker alternative. This convention had been around since the teens. Recalling his days shooting comedies at Essanay, Arthur Miller wrote, "My first instructions were that the faces of the actors must be white and the sets fully lit with no shadows. The 'writer' was around all the time when we were shooting our first comedy of the series for Essanay, constantly reminding me that comedy couldn't be played in the dark. He said it over and over, as if to make it perfectly clear that he was a

Figure 3.13 Softer lighting for the Greta Garbo romance *The Mysterious Lady* (1928).

Figure 3.14 The lighting illuminates the entire space for the Buster Keaton comedy *Sherlock, Jr.* (1924).

genius."[46] Miller did not enjoy shooting in this style, but it was justified by the demand for a bright mood, combined with a need for maximum clarity. Similarly, the comic mystery plot of Buster Keaton's *Sherlock, Jr.* (1924) could have allowed the use of shadows and contrasts, but cinematographers Byron Houck and Elgin Lessley wisely decided to keep the action bright (fig. 3.14). This accomplishes two functions: first, it maintains a positive mood; second, it allows us to see the action clearly—an essential advantage for a comedian like Keaton, whose gags often involves the creative use of space.

By contrast, dramatic films like *The Wind* and *The Conquering Power* (Rex Ingram, 1923) often employ darkness during the most serious scenes. As in the melodrama, cinematographers sometimes combine the conventions of dramatic lighting with the conventions of effect-lighting. In one of the most serious scenes in the romantic drama *Flesh and the Devil* (Clarence Brown, 1926), Leo (John Gilbert) attempts to strangle his beloved Felicitas (Greta Garbo). Cinematographer William Daniels sets the appropriate mood by keeping the foreground dark, motivating the technique as an effect by placing a table lamp behind Leo. Belasco himself would have been proud of such work.

The theater and the cinema are both drawing here on a long tradition in Western art, associating shadows with serious subjects. In Rembrandt's etching of the *Entombment* (1654), the darkness is so deep that it nearly swallows the figures. In Jacques-Louis David's *The Lictors Bring to Brutus the Bodies of His Sons* (1789), the women mourn in bright sunlight, but Brutus himself receives the news in shadow, the darkness becoming an outward signifier of gravitas. In Eugène Delacroix's *Pietá* (1850), a shadow envelops the twisted form of Jesus. Homage to such traditions reinforced the idea that the cinematographer was a painter with light.

Sometimes, the conventions of effect-lighting and genre/scene lighting would blur together. Consider the following passage from an *AC* article by Virgil Miller, who co-photographed the expressive horror film *The Phantom of the Opera* (Rupert Julian, 1925):

> An underworld setting can not be lighted like a ball-room. In the underworld den we play for the weird, shadowy, suggestive effects, lights from beneath suggestive of infernal fires, feeding the imagination, and breathing that into the picture that makes the observer feel as well as see the story. In the ball-room we look for brilliant overhead or "face level" lightings, suggesting cheerfulness and freedom from the shady things in life.[47]

Miller's examples can be taken in two ways. On the one hand, they seem to be examples of effect-lighting, since he is trying to imitate the appearance of light in a particular location. On the other hand, they seem to be examples of the genre/scene conventions: by using the lighting techniques to evoke certain associations, he hopes to establish the appropriate mood for the story. The word "atmosphere" often carries this complex meaning: it can refer to the atmosphere of a location, to the atmosphere of a story, or to an atmosphere that suits both of them at the same time.

Does this mean that we have one set of conventions here, not two? Although we could simplify matters by calling these rules the "atmosphere" conventions, I think it is a good idea to think of the effect-lighting conventions and the genre/scene conventions as two distinct sets of rules that could be combined in practice. Even in a comedy, a cinematographer might need to produce some dark shadows, to convey the impression of an alley or a basement. Even in a drama, a cinematographer might need to employ bright lighting, to convey the impression of a morning or an afternoon. Similarly, a cinematographer might use soft or hard lighting to suit the mood of a romance or melodrama, without necessarily seeking to motivate the technique as a particular effect.

A final troubling issue concerns the artistic status of these conventions. It is tempting to say that cinematographers used mood-lighting techniques for artistic purposes, while using effect-lighting to motivate or conceal those techniques. The problem is that this assumes that effect-lighting itself had no aesthetic value, apart from its value as a method of concealment. However, imitation has been an important component of the definition of art for centuries. We should expect artistically ambitious cinematographers to have placed a high value on accurate detail, regardless of expressive considerations. Even before the formation of the ASC, Tony Gaudio complained that the obligation to light the actors might compromise the art of imitation:

> Only in a night scene can the real lighting be shown. In this the character is illumined presumably from moonlight streaming through the windows, or from a fireplace or from a table lamp. In these scenes, we can throw our artificial light where the natural light would fall, and the characters are illumined in a realistic manner.
>
> The motion picture photographer is decidedly up against it. The necessities of the drama demand that the characters be shown plainly to get in all of their action. If the natural lighting is used, the face of the character is lost and the action is sacrificed. So the photographer must sacrifice naturalness and the artistry of reality for the benefit of playing up the faces.[48]

Gaudio values effect-lighting not because it is expressive, but because it is realistic. Accurate detailing is an artistic end in itself, and he hopes for a day when the cinematographer will be able to photograph a scene using natural light and nothing else.

◎ ◎ ◎

The effect-lighting and genre/scene conventions were two more steps in the ASC's program to develop the art of Hollywood lighting. Instead of working with one neutral norm, cinematographers embraced the ideal of differentiation, resulting in several distinct practical norms. The cinematographer who understood these rules was an efficient artist who could harness the drama of light.

Organizing the Image

Composition involves the arrangement of distinct pictorial components into a larger whole. This definition sounds simple enough, but
it becomes more complicated when we think of composition conventions in two different ways. First, composition involves the abstract arrangement of tonalities on the two-dimensional screen. A filmmaker
might balance a highlight on the left with a highlight on the right, or intensify a highlight by placing it next to the deepest shadow. Second, composition is a representational problem. To compose a picture is to compose a picture *of* something: typically, a three-dimensional space occupied
by human figures engaging in some sort of narrative action. Composition
can aid representation by rendering the space easy to read and the story
easy to follow. In general, cinematographers favored the second sense of
the term "composition"; however, their devotion to the ideal of pictorial
balance suggests that they never completely abandoned the first.

For instance, consider the most basic Hollywood composition convention: put the brightest light on the most important story point. Figure 4.1
is from *The Rag Man* (Edward Cline, 1925). Here cinematographers Frank
Good and Robert Martin employ this simple convention in a very efficient
way, guiding our attention to the center of the image, while keeping our
attention away from the relatively insignificant background. The lighting
serves the story by making it easy to follow. At the same time, the filmmakers maintain a certain level of compositional balance. The centering
may arrest our attention, but it also creates a symmetrical image, with
darkness at the margins and light in the middle.

Figure 4.2, from Frances Marion's *The Love Light* (1921, photographed
by Charles Rosher and Henry Cronjager), offers a more complicated example. Here there are two points of interest. Fittingly, there are two major

Figure 4.1 Lighting draws our attention to the center of the composition in *The Rag Man* (1925).

areas of light, one for each character. This image even includes a subtle hierarchy: Mary Pickford receives more light than Fred Thomson does, indicating that her character is more significant. Also notice that the upper right section of the back wall has been given more light than the needs of storytelling would require, thereby balancing the area of brightness in the lower left corner. This enhances the sense of pictorial order without sacrificing storytelling in any way. As always, lighting is multifunctional.

Even these simple examples can be perceived as attempts to elevate cinematography to the level of art. The use of composition to guide the eye is a principle borrowed from the art of painting. The principle was simple enough, but its application required a certain amount of interpretation. Some scenes might be better served by a strong hierarchy, while other scenes could require gentle gradations of emphasis. This is why cinematographers saw composition as a task fit for an artist, not a mechanic.

In practice, many cinematographers relied on a simple guideline: when in doubt, put a little extra light on the foreground. Since the important story point was usually in the foreground, this conventional wisdom would allow the cinematographer to direct attention to the story without a lot of effort. Still, even this rule required some consideration: the cinematographer needed to decide how to differentiate foreground and background in terms of contrast. When should the contrast between foreground and background be strong? When should it be subtle?

The principle of contrast was particularly important to the illusion of roundness. As we saw in the previous chapter, ambitious directors like Cecil B. DeMille experimented with high-contrast lighting effects in the mid-teens. Throughout the twenties, however, DeMille adopted a more

Figure 4.2   Multiple points of interest in *The Love Light* (1921).

moderate approach, favoring relatively low-contrast imagery. In his auto-biography, DeMille explains this shift in representational terms:

> Alvin Wyckoff, the cameraman on most of my early pictures, loved strong light. He was a skillful cameraman and on most camera angles and effects we agreed completely, but it was a struggle to convince him that there are shadows in nature and that mingled light and shade can be more beautiful than glare. When Alvin got the point, he got it strong. He coined a word to describe me: I was the director who liked everything "contrasty." When that reputation spread among the cameraman, of course, I had another struggle to make it clear that contrast of light and shadow meant what you naturally see with normal eyes if you look out your window in the late afternoon when the long shadows of the trees fall across the lawn, and not the solid pools of Stygian black in the middle of a brightly-lighted scene which some cameramen began to create in order to please this director who wanted it all "contrasty."[1]

DeMille favors a moderate option, between one extreme—giving the entire image an equal amount of illumination, which tends to flatten the image into a plane of undifferentiated highlights—and another—creating intense contrasts, which tend to make backgrounds disappear into undifferentiated shadows. By producing gentler gradations between highlights and shadows, the cinematographer can produce a more compelling impression of depth.

Figures 4.3, 4.4, and 4.5 represent some of these compositional alternatives. In 4.3, an introductory shot from *The Cheat* (1915), strong general illumination flattens the space into an abstract pattern. Figure 4.4 is an example of effect-lighting from *The Golden Chance* (1915). The lighting casts unpleasant shadows on the face of the female protagonist, and the

Figure 4.3 Bright illumination produces weak modeling in an introductory shot from *The Cheat* (1915).

Figure 4.4 The darkness of the shadows produces a loss of detail in the background, in *The Golden Chance* (1915).

Figure 4.5 *A Romance of the Redwoods* (1917): Key and fill work together to model Pickford's face.

flat white background has become a flat black plane. In 4.5, from *A Romance of the Redwoods* (1917), DeMille and Wyckoff have found a more moderate solution. The background is light enough to create a sense of depth without becoming a distraction. Meanwhile, the lighting models Pickford's face with a carefully balanced blend of highlights and shadows.

To our eyes, the moderate solution of *A Romance of the Redwoods* may appear less striking than the bold effect from *The Golden Chance*. For this reason, it is tempting to see this development as an abdication of artistry in the face of classicism's demand for unobtrusive craftsmanship. However, DeMille does not argue that the shift in style diminished the artistry of his works. If anything, he sees this move as a step forward in artistic accomplishment, capturing the inherent beauty of the world as it is seen with natural vision. Of course, we should be careful about taking DeMille's rhetoric of artistry too seriously. James Wong Howe was working for DeMille as an assistant around this time, and he tells a different story: "DeMille was very commercial. I remember he used to tell Alvin Wyckoff, 'Look, I spent a lot of money on this set. I want to see every corner of it.' He was a showman."[2] DeMille may have talked a lot about Rembrandt in interviews, but he was always looking for ways to emphasize spectacle.

DeMille does not give any technical details on the development of this more moderate approach, but a 1926 article by Wyckoff himself offers some clues. Wyckoff writes:

> With the advancement in perfection of the Cooper-Hewitt, it was necessary to introduce a method whereby light could be projected through the illumination of the Cooper-Hewitt light that would "round out" the scene with degrees of contrast, so controlled that the effect obtained would render a natural impression. For this purpose the spot light was adopted, ranging in power from the tiny arc with a carbon no longer than a small pencil, generally used to portray a lighted candle, to the immense and powerful searchlight that would produce the effect of bright rays of sunlight. With this perfection it has become possible to project scenes upon the screen that are bewildering in their natural fidelity.[3]

As Barry Salt has described, Cooper-Hewitts were large tubes that provided a general soft light, similar to diffused daylight. The problem was that relying exclusively on Cooper-Hewitts ran the risk of creating a flat image, since the lights were too soft to produce a noticeable shadow. The arc light, meanwhile, provided a much stronger, harder light, but it introduced a different kind of risk: the image might be too contrasty to allow for a range of gradations. For this reason, Wyckoff recommends the use of both tools. An arc light could create the shadows necessary for the illusion of roundness, while the Cooper-Hewitts would provide an overall fill level to prevent the shadows from becoming too dark.[4]

The composition conventions were more generalized than the rules of figure-lighting, genre/scene lighting, and effect-lighting. As such, they often incorporated elements of the other conventions. Wyckoff was known for his effect-lighting, and he treats the spot light as a potential effect, capable of imitating candlelight, sunlight, or anything in between. The same spot light could also function as the key-light of a three-point lighting arrangement.

Still, the distinction between highlights and general illumination is not necessarily the same thing as the distinction between the key-light and the fill light. Many cinematographers would begin the process of lighting by creating a base level of ambient light that would spread over the entire set. This technique would ensure a minimum level of exposure for every detail in the shot. Producing the general level of illumination would require several different lighting instruments, such as the Cooper-Hewitts mentioned by Wyckoff. After establishing this base level, a cinematographer could introduce other lights, such as the key-light and back-light, which are familiar components of the three-point lighting system. For fill light on the performer, the cinematographer had three choices. He could simply rely on the general illumination to provide the fill. In this case, the shadows on the actor's face would be just as dark as the shadows everywhere else on the set. If the cinematographer wanted the shadows on the face to be brighter than the shadows on the set, then he would need to bring in an additional spotlight to provide the fill. He could use black pieces of cloth to produce "negative fill," enhancing the shadows on the fill side by cutting off some of the ambient light. In order to keep these distinctions clear, it might be useful to think of "key-light" and "fill light" as terms of figure-lighting, while using "highlights" and "general illumination" as terms of composition, even if their functions are sometimes performed by the same instruments.

Not surprisingly, we can find a theatrical precedent for the recommendations of DeMille and Wyckoff. In the late nineteenth century, Adolphe Appia had described his distinction between general illumination and focused lighting in the theater: "Part of the lighting equipment will be used for general illumination, while the rest will cast shadows by means of exactly focused beams. We shall call them 'diffused light' and 'living light.'"[5] Eventually this distinction became a staple of theater lighting.[6]

In addition, the tradition of painting provided an important influence. Some filmmakers even went so far as to imitate certain painterly techniques. For instance, the director Maurice Tourneur was known for using a painterly technique known as the *repoussoir*. Figure 4.6 provides an example, from *Lorna Doone* (1922, photographed by Henry Sharp). Rather than create separation by lighting the foreground more than the background, Tourneur reverses the technique, casting the foreground in shadow. A 1923 article in the *Transactions of the Society of Motion Picture Engineers* suggests that the artist Gustave Doré may have been an inspiration for this technique: "The picture may be given apparent depth, by lighting the background more intensely than the foreground, a familiar trick of Gustave Doré, who seems to have imagined cinematic lighting in the days of the zoetrope."[7] Looking farther back, we can note that this technique had long been associated with French Academic painting. Although the *repoussoir* is intended to function as a sign of Tourneur's artistic sophistication, it soon became a thoroughly standardized technique.

Whether using the basic conventions of contrast or the more specialized convention of the *repoussoir*, the cinematographer was guided by an ideal of illusionism—another ideal with an ambiguous meaning. DeMille and Wyckoff seem to refer to the illusion of presence: notice how DeMille praises natural vision as an ideal, while Wyckoff suggests that images will have a "bewildering" power. There is evidence for a different interpretation in a 1917 article by Wilfred Buckland, the theatrically trained set designer who was responsible for some of the lighting effects in DeMille's early films. Buckland rejects absolute realism as an ideal, and proposes the use of the evocative paintings of Whistler and Corot as models.[8] Composition can produce a painterly illusion of roundness, without necessarily creating a strong illusion of presence. The spectator

Figure 4.6 A darkened foreground helps create a sense of depth in *Lorna Doone* (1922).

should see depth in the image, while retaining a subtle awareness of the fact that artful lighting has produced this depth.

Moderate contrast soon became the norm for cinematic composition, but this still allowed room for variation. The cinematographer was supposed to avoid the extremes of absolute flatness and utter darkness, but he was free to explore the vast range of options located between them. Indeed, as we have already seen, the genre/scene conventions encouraged the cinematographer to do just that. By dimming the general lights, the cinematographer could create the strong contrasts demanded by the melodrama. By increasing the general lights, he could create the bright atmosphere of the comedy. Either way, he simply had to avoid the extremes. A brightly lit comedy should have enough shadows to achieve an appropriate amount of modeling, and a contrasty melodrama should have enough general light to preserve some detail in the shadows.

The conventions of effect-lighting also encouraged a certain amount of variation. As cinematographer Charles Clarke once explained: "Regardless of whether one does full light scenes or extreme effect scenes, the key-light intensity remains the same for any given film speed and lens stop. Only the amount of fill light is changed to create these effects. The first is a mechanical requirement for the proper exposure of the film—the second is up to the artistry of the cinematographer."[9] For instance, to produce a nighttime effect, a cinematographer could keep the key-light the same, reducing the fill light to increase the contrast. More fill light would produce a daytime effect. These conventions gave the cinematographer a range of artistic options.

Although high-contrast compositions tend to be more striking, we should remember that the cinematographers of the silent period took special pride in their ability to craft low-contrast imagery. Indeed, a major trend throughout the twenties, known as the "soft style," encouraged cinematographers to produce the gentlest possible gradations. Kristin Thompson has examined the development of the soft style in detail.[10] Just as pictorialist photographers had softened their pictures to make them look more like paintings, Hollywood cinematographers began to soften their images to resemble artfully crafted photographs. The trend encouraged a variety of cinematographic techniques that involved lighting, lens choice, and developing. Some of the most popular techniques were as follows.

1. Placing diffusion on a lamp would make the source of light larger, thereby eliminating the hard lines differentiating highlight from shadow.

This corresponds to the modern usage of the term "soft light." Here, the "softness" refers to the edges of the shadows.

2. The cinematographer could produce low-contrast lighting by reducing the difference between key and fill, or the difference between highlights and general illumination. The result would be a different kind of "soft" image—an image with gentle gradations of gray. This option could be accomplished with or without the use of lamp diffusion.

3. Another way to manipulate contrast was to overexpose the negative and reduce development time. This would eliminate the darker shades of black, while favoring subtle shades of grey.

4. Using less light allowed the cinematographer to open the aperture, thereby reducing the depth of field. Here, the "softness" would refer to the background, which would be rendered in soft focus. In exterior situations, when it was not possible to reduce the light level, a cinematographer might use a neutral-density filter to achieve the same effect.

5. A few ambitious cinematographers stretched sheets of muslin behind the actors, reducing the background's contrast without throwing it out of focus.

6. By placing a sheet of gauze over the lens, a cinematographer could make the entire image appear in soft focus. Some cinematographers liked to burn a hole into the center of the gauze. This would soften the edges of the image while maintaining a sharp center. Note that this use of diffusion (lens diffusion) is very different from the lamp diffusion of the first option above. Lens diffusion was essentially a camera technique, though it could influence the lighting scheme by making the highlights appear to glow.

At first the soft style was seen as a mark of artistic pretension. But by the mid-twenties several soft style techniques had become basic standards. Cinematographer Stephen S. Norton explains some of the more common conventions:

> The effect caused by gauze helps to tone down the tempo of the scene in many cases when applied to interior scenes of a pathetic nature, but gauze cannot be exclusively depended on for this effect, as the lightings must also be of the subdued type.
>
> Another pleasing use for the soft focus is in exterior garden or rural locations where there is an opportunity for fine composition and soft lighting effects and where the sequence is of a romantic nature.

When applied to close-ups gauze is very often of great importance, especially where the star or other member of the cast are in front of a large group of people and the action is intended to make a general impression on the audience. In this case the background is softened or subdued while the star is brought out in clear contrast.

There are many times when the soft effects prove to distract rather than to help and one of several instances is when used in sequences of a melodramatic nature where the action is tense and fast and where the surroundings correspond. Usually these scenes are photographed clear and sharp with lots of snap. To cut in a close-up of soft focus many times distracts or jars and tends to slow the action up.[11]

Norton argues that gauze should not be used all the time, no matter how beautiful the effect. Instead, the cinematographer should use gauze only when it is appropriate for the story. Gauze can enhance the atmosphere for a "pathetic" or "romantic" scene, or it can direct the audience's attention to the star in a close-up shot. By contrast, pictorial beauty is out of place in a melodrama, a genre that favors sharpness. It is tempting to read this passage as an attempt to limit the influence of the soft style, confining it within the boundaries set by the classical norms. However, this interpretation misunderstands the role of the ASC. Far from attempting to limit the style, the ASC is attempting to expand the style's influence. It is likely that articles like this one helped popularize the technique, precisely by specifying the contexts in which it was most useful. Because the ASC had developed an eclectic discourse of art, it favored multifunctional solutions. Ideally, a technique would enhance pictorial beauty, establish the mood, and direct the spectator's attention at the same time. In other words, cinematographers like Norton do not reject overtly pretty pictures because they are too artistic; they reject them because pretty pictures are not artistic enough. A pretty picture lacks the multifunctionalism that the ASC taught its cinematographers to prize.

This raises a challenging problem for the analysis of Hollywood's composition conventions. Invisibility is often cited as a core ideal of the Hollywood style. At times in the early years it appears that cinematographers had already adopted this ideal, favoring unobtrusive storytelling over flamboyant technique. However, it is not hard to find examples of films that flaunted their pictorial beauty, such as Griffith's *Broken Blossoms* (1919, shot by Billy Bitzer) and Murnau's *Sunrise* (1927, shot by Charles Rosher and Karl Struss). One solution would be to say that invisibility was the norm, while acknowledging that various exceptions to it were allowed, such as the masterpieces of the soft style. However, when we look at the discourse of cinematography, we find that things are much more complicated. For instance, Hal Mohr, who co-photographed the soft style masterpiece *Sparrows* (William Beaudine, 1926), once claimed that the techniques of diffusion were supposed to enhance the illusion of presence; it was films with sharp edges that reminded audiences they were looking at pictures.[12]

The norm-and-exceptions model does not do justice to the complexity of Hollywood lighting during the silent period. Most cinematographers were fully committed to the ideal of storytelling, but they did not always assume that stylized techniques contradicted this principle. In the introduction I argued that a close look at cinematographic ideals may force us to revise our understanding of the principles of classicism, taking note of some surprising nuances and complications. To get a better sense of this complexity, I would like to examine the discourse of three distinct cinematographers: John F. Seitz, Walter Lundin, and L. Guy Wilky. We will see that each one offered a slightly different account of the relationship between storytelling and invisibility.

◎ ◎ ◎

Throughout the twenties, John F. Seitz was one of the most widely respected cinematographers in the world. Working with director Rex Ingram, Seitz had photographed such beautiful films as *The Conquering Power* and *The Four Horsemen of the Apocalypse* (both 1921). He would later bring his talents to Paramount, where he shot such classics as *Sullivan's Travels* (Preston Sturges, 1941) and *Double Indemnity* (Billy Wilder, 1944). In 1923 Seitz contributed an article to *American Cinematographer* in which he maintained that experience, not academic schooling, is the best way to learn cinematography. In this article Seitz suggests that there are four different kinds of cinematographer: newsreel, travelogue, comedy, and drama. Newsreel cinematographers must be "go-getters," while travelogue cinematographers must have a "love of nature." As for the comedy cinematographer, Seitz writes that his goal is simply to "obtain or present motion in the most effective way for his purpose." To this end, the comedy cinematographer favors lighting schemes that are "simplified" and "unobtrusive."[13] Seitz makes no mention of artistry in regard to the first three categories, which in his view are primarily concerned with accurate mechanical reproduction, not creative interpretation.

Seitz switches to the rhetoric of artistry when he gets to his own category: dramatic cinematography. He writes:

> The cinematographer in the dramatic field is more of a photographer and less of a cinematographer than the comedy cameraman, his action contains less of the physical and more of the mental, consequently he is less concerned with motion and more with lighting and tone.
>
> In the best of the dramatic productions we often see examples of what is at once the science and art of cinematography—the perfect harmony of the photography with the mood of the story and of the players. This effect, when obtained, approximates perfection, as we now understand it. It is manifest that this perfection can only be attained by men who, through long, patient experience, have gained that fine sensitiveness so necessary to produce the exact tone and quality needed, and this is cinematographic art, for, after all, art is not a thing, but a quality.[14]

Seitz has taken the mechanics-to-artists narrative of the ASC and mapped it onto an openly hierarchical theory of genre. Whereas he aligns nondramatic cinematographers with manual labor, the dramatic cinematographer works with ideas and emotions. The dramatist's sensitivity will allow him to interpret the mood of the story intelligently, and his skill will allow him to craft the precise visual style to suit the needs of the story. He is at the top of his field, and he deserves to be treated like a professional artist.

Seitz is proposing a different way of thinking about the relationship between style and story. Significantly, Seitz does not need to choose between art and storytelling. This is because he defines storytelling in an expressive way. The story is more than just a causal chain of events that runs through the entire film; it is a generator of dramatic, emotional moments. It is these emotional moments that call for artful cinematography. Because Seitz defines artistry as the creation of story-specific moods, he has every reason to embrace the narrativization of cinematography. More story means more mood, and more mood means more opportunity for artistry. The cinematographer who fancies himself an artist will look upon the dramatic film as the perfect forum for his talents.

Does this mean that Seitz thought his style was invisible, because it was subordinate to story? Or did he think his style was highly visible, because it was so expressive? I am not sure that Seitz himself would have phrased the problem in this particular way. We have seen that he describes comedy cinematography as "unobtrusive." Because he is drawing a contrast between comedy cinematography and drama cinematography, the implication is that Seitz thinks of his own work as being much more openly artistic. At the same time, he suggests that the artistry arises from a commitment to the story. In other words, Seitz does not make the assumption that unobtrusiveness conflicts with storytelling. He does not see a contradiction here—he sees convergence. Seitz's artistic accomplishment is manifest in the careful harmony of story and style. His art is not subordinate to storytelling. His art *is* storytelling.

It is tempting to suspect that Seitz's views do not represent the mainstream of cinematographic discourse. After all, he worked for Rex Ingram, perhaps the only man in Hollywood who could claim to have received training as a sculptor at the Yale University School of Art. Ingram and Seitz later left Hollywood for Europe, where they continued to make extravagantly aesthetic productions. Perhaps this move across the ocean indicates that these two figures had been working at the margins of the Hollywood style all along.

However, there is a marked difference between being at the margins and being at the top. In the eyes of the ASC, Seitz was not a marginal figure. He was a model to be emulated. Indeed, Seitz became president of the ASC a few years later. While it is true that not every Hollywood film looks like a Seitz film, this should not distract us from the fact that many important filmmakers may have endorsed Seitz's general philosophy regarding the artistry of dramatic cinematography. Indeed, it is not an exaggeration to say that the silent studio period was a high point of cinematic pictorialism. In addition to Ingram, other directors with pictorially ambitious styles include Maurice Tourneur, Clarence Brown, D. W. Griffith, Josef von Sternberg, Erich von Stroheim, John Ford, and George Fitzmaurice. Among cinematographers, we might list Charles Rosher, Arthur Edeson, Charles van Enger, Joseph August, William Daniels, Hal Mohr, Arthur C. Miller, and Victor Milner.

We might suspect that comedy cinematographers would reject Seitz's ideas, given that Seitz had denigrated them as second-class citizens. However, a 1924 *American Cinematographer* article by Harold Lloyd's cinematographer Walter Lundin shows that the group's response was much more complicated. As a slapstick cameraman, Lundin already had an unusual relationship with the "classical" norms. Following the lead of Tom Gunning, contemporary film historians have discussed the slapstick comedy in terms of the transition from the cinema of attractions to the cinema of narrative integration.[15] Whereas the cinema of attractions offers a nonlinear series of ephemeral shocks and delights, the cinema of narrative integration subordinates attractions to a linear causal chain. The slapstick comedy is often seen as a historical anomaly, preserving the cinema of attractions mode long after the cinema of narrative integration had become dominant. Harold Lloyd's films are positioned between the two poles, combining a high degree of narrative integration with a generous supply of attractions.

Lundin's article provides some support for this analysis of Lloyd's films, but it also suggests the need for a few subtle revisions. In a section titled "Story Subordinate," Lundin writes:

> In the olden days, comedies, I might say, were objective to an extremity. All action, never a dull moment, keep the audience on the edge of the chair, story and plot always subordinate to gags.
>
> Comedies must still have their gags, but even therewith this medium of motion picture entertainment is no longer identified with action at any cost—and there is still plenty of action—but has, on the contrary I might again hazard an opinion, begun to stroll the paths of the subjective. By that I mean that comedies of the outstanding class are no longer a series of incoherent situations which, though laughable, were not always quite reasonable.
>
> [*New section*] "Story Carried Throughout"
>
> No, the feature-length comedy has changed this. There is a thread of story that runs through the channel of humor; there are drama and moments of pathos in the most hilarious of comedies—and all this directly affects the cinematographer who films such productions.[16]

In modern terms, Lundin is discussing the slapstick comedy's belated transition from the cinema of attractions to the cinema of narrative integration. The early slapstick comedies were once uncoordinated strings of gags, but now a coherent narrative works to integrate the gags into a larger whole. This supports the standard reading of Lloyd's films, but a few details call for more nuance. Gunning argues that the cinema of narrative integration works to contain the attractions. According to this account, the primary function of narrative is to serve as a system of organization, subordinating the attractions to the causal chain. Gunning adds that this system is never completely successful. Even within the cinema of narrative integration, attractions will occasionally appear to challenge the dominance of the guiding narrative. While it is true that Lundin theorizes narrative as a system of organization, he does not suggest that organization is the primary function of the narrative. Instead he argues that narrative offers emotional appeals of its own, in the form of moments of pathos. This has important consequences for our understanding of the relationship between the narrative and the attractions. Because narrative and gags both offer emotional appeals, Lundin does not need to suggest that the two systems compete for dominance. The old slapstick film may have subordinated story to gags, but the new slapstick film does not *reverse* the terms. The new slapstick film *combines* the two for maximum emotional appeal.

In addition to proposing a different way of thinking about the narrative, Lundin hints at a different way of understanding the invisible style. According to Gunning, the cinema of attractions dazzles spectators with moments of pictorial display, while the cinema of narrative integration employs a more invisible style to encourage the experience of absorption. By contrast, Lundin writes:

> But the majority of that small minority of motion picture patrons who have ever recognized photography in the least, always have been impressed with something "beautiful"—such as lovers under the blossoming trees in springtime, etc., etc. They may have a faint idea that comedy cinematography entails danger of life and limb as well as a knowledge of the most intricate details of the camera, but even with this suspicion they are never able to place it on a plane of comparison with the dramatic.
>
> The feature-length comedy, however, with its plot, its recognition of the subjective, its points of pathos and drama, has changed the outlook of the cinematographer making the same. He is no longer consigned to the oblivion of what is considered as ordinary, but is given the opportunity to step forth with sequences, the photography in which vies with that in dramas for pictorial beauty that arrests the attention of the critically inclined.[17]

According to Lundin, it is the undernarrativized slapstick comedy that produces an invisible style, as the cinematographer is forced to subordinate his considerable technical skills to the needs of the gag. By contrast, it is the narrativized feature-length comedy that gives the cinematographer the opportunity to "step forth" with images of eye-catching artistry. At first blush, this seems paradoxical, because we expect the nonnarrativized film to offer itself as a display, and the narrativized film to render style invisible. Here narrativization allows *more* display, not less. This seemingly paradoxical idea becomes quite plausible if we suppose that Lundin is drawing on the same ideas as Seitz. Strong emotional situations call for strong images to express those emotions, as when images of blossoming trees in springtime set the mood for romance. By providing a variety of emotional situations, narrative gives the cinematographer more opportunities to practice his art.

Seitz and Lundin both appeal to a theory discussed in the previous chapter: the right-mood-for-the-story theory. In both cases the theory inspires the cinematographers to reject the unobtrusiveness that we normally associate with the classical Hollywood cinema. However, we should remember that the discourse of the ASC was highly varied, especially during its formative years. An article by L. Guy Wilky illustrates this signature eclecticism. On the one hand, Wilky openly endorses the ideal of invisibility; on the other, he contends that the invisible style is appropriate for some films, but not for others. Specifically, he proposes three categories of cinematography:

> The cinematographer who has a theme of rousing action with which to work—costume stuff, with plenty of swordplay and backgrounds of castles, and the like—possesses the opportunity to blossom forth with the kind of motion photography which, if properly done, must command the attention of even the most casual layman. . . .
>
> On the other extreme, we encounter comedy cinematography, replete with "special effects," necessary in aiding and abetting the spontaneous registering of the endless "gags" on which the average short comedy thrives. . . . The work of the comedy cinematographer, in short, is such that it, too, stands out for recognition to all those who view motion pictures.
>
> Between the foregoing two extremes then, there lies a field of cinematography wherein the highest compliment that could be paid to the cinematographers, who are giving forth their efforts in it, is that their work, in a given production, is scarcely "noticeable."[18]

This first category might include Seitz's work on swashbucklers like *The Prisoner of Zenda* (1923). The second category might include Lundin's work on slapsticks like *Safety Last* (Fred Newmayer and Sam Taylor, 1923). Wilky casts these categories a bit differently than Seitz or Lundin would. Whereas Seitz defines artistry in terms of mood, Wilky proposes that the artistry of the action film is the art of pictorial beauty. Whereas Lundin appears almost embarrassed by his unaestheticized straight comedy work, Wilky claims that comedy cinematography gives spectators the chance to admire conspicuous displays of technical skill. In both cases, Wilky suggests that cinematography works by attracting the attention of the audience—even at the cost of luring their attention away from the story.

Wilky's third category would include the widest range of films, including his own work for Cecil B. DeMille's brother, William de Mille. Unlike Seitz and Lundin, Wilky champions unobtrusiveness as an artistic ideal. To complicate the matter further, Wilky embraces invisibility while employing the same right-mood-for-the-story theory that is endorsed by Seitz and Lundin. Referring to his own films, Wilky offers several examples: gloomy lighting for a crime film, soft photography for a romance, bright lighting for a comedy. Cinematography becomes invisible when the style is in perfect harmony with the story.

Theorists of the classical style sometimes suggest that filmmakers valued invisibility as an end in itself. Alternatively, some suggest that the invisible style was designed to increase the illusionism of the classical cinema. A spectator who does not notice the style becomes absorbed into the world of the film, experiencing that world as if from the position of an ideal observer.[19] Neither explanation applies to this case. Wilky does not value an invisible style for the sake of invisibility, nor does he value invisibility because it helps to produce an illusion of seeing the fictional world from an idealized position. Instead, he values invisibility as an index of his work's sense of balance. Spectators become acutely aware of style when it is discordant with the story. When spectators ignore style, it is proof that the cinematographer has achieved the desired harmony. Wilky departs from Seitz and Lundin on this point. Seitz and Lundin aspire to create harmony between style and story, but they suggest that spectators might appreciate the resulting harmony as an attraction in its own right.

In other words, we should resist the temptation to assume that at this point cinematographers have already adopted all of the concepts that we now associate with classicism: the commitment to storytelling, the invisibility of style, and the illusion of presence. Theories of the classical cinema often treat these concepts as a trio of related ideas: the dominance of narration makes style invisible, which facilitates the illusion of presence. The ASC did not link these three ideas together, and a variety of views could be collected from the association's members. Seitz, Lundin, and Wilky are all committed to storytelling. They take three different positions on the virtues of invisibility in which the ideal observer is nowhere to be found.

Furthermore, one can endorse the ideal of storytelling without adopting the persona of the humble craftsman. The skillful cinematographer has several kinds of aesthetic knowledge. First, he must interpret the story correctly, identifying when a film shifts from comedy to drama, or from romance to tragedy. Second, he must know what style is appropriate for each particular scene: when to be light, when to be dark, when to be soft, when to be sharp. This presupposes a third ideal: the ability to master a wide range of techniques. Finally, the cinematographer dexterously combines all these ideals together to work toward an admirably aesthetic goal: the production of emotion.

◎ ◎ ◎

I would like to conclude this chapter by considering some examples from the cinematography of Walter Lundin. My intention is not to elevate Lundin to the pantheon of great cinematographers, though I certainly think that his work is underrated. Rather, I hope to demonstrate the power of the rhetoric of light by showing how it influenced the work of a skillful but hardly unusual Hollywood cinematographer. It would be too easy to point to a Tourneur film or a von Sternberg film to prove the point that Hollywood tolerated and even encouraged a certain degree of pictorial artistry during the silent period. The argument is more convincing when we find the same level of pictorial ambition in the work of a cinematographer whose primary task was photographing the amusing gags and sentimental plots of a Harold Lloyd film.

The films of Lundin employ all four of the conventions we have considered so far: figure-lighting, effect-lighting, genre/scene lighting, and compositional lighting. Lundin is particularly good at making subtle adjustments to his lighting schemes to allow them to serve several functions at the same time. For instance, consider two different images of Jobyna Ralston: figure 4.7, from *Hot Water* (Newmeyer and Taylor, 1924), and figure 4.8, from *Girl Shy* (Newmeyer and Taylor, 1924). The first example follows the rules of three-point lighting, with a key-, fill, and backlight. In keeping with the composition conventions, the face is a bit brighter than the background, directing our attention while producing a sense of roundness. At the same time, the overall tonality is bright, setting a suitable mood for a broadly comic scene. The second example combines three-point lighting with two other strategies. First, the background is considerably darker than the foreground. Second, Lundin has added diffusion to the edges of the image, while keeping the center sharp. These two strategies serve a variety of functions. *Girl Shy* is a much more romantic film than *Hot Water*, and the soft technique sets the right mood for romance. The combination of softness and darkness produces an equally appropriate mood of pathos. Meanwhile, the diffusion and the

Figure 4.7 A simple three-point lighting scheme for the broad comedy of *Hot Water* (1924).

Figure 4.8 A more glamorous technique for the romantic comedy *Girl Shy* (1924).

lighting keep our attention on the foreground—an important point, because the busy background might have created a distraction. Finally, the image has a pictorial beauty that can be appreciated for its own sake. Note that this is not necessarily a case of style challenging the dominance of story. As Lundin himself would argue, the dramatic nature of the scene gives him the opportunity to showcase his skills.

Many of Lloyd's films contain elements of suspense. Lundin consistently takes advantage of these opportunities to produce images with strong contrasts, following the conventions of effect-lighting and genre/scene lighting. In *Hot Water* Lloyd's character mistakenly believes that he has killed his mother-in-law. Worried that he will soon be arrested, he walks into another room—only to find himself "caught" in a jail-like pattern (see fig. 4.9). Here, the effect-lighting follows (and gently mocks) the genre conventions of the melodrama. This example supports Lundin's argument about the advantages of narrativization. Far from containing the effect, the narrative encourages the spectator to notice the lighting. A

Figure 4.9 Harold Lloyd walks through a pattern of bars in *Hot Water*.

Figure 4.10 Simple exterior lighting for a gag in *Girl Shy*.

spectator who fails to understand the narrative—or fails to look at the lighting—will fail to get the joke. This shot was probably tinted blue in the 1924 print, making the shot even more striking, while increasing the plausibility of the nighttime effect. Story and spectacle work together to produce a multifunctional result.

Of course, Lundin also knows how to employ an unobtrusive style when appropriate. In *Girl Shy* (fig. 4.10), Lundin refrains from adding any flourishes for an outdoor scene when Lloyd is hanging onto the back of a moving car. Lundin himself may have been a little embarrassed about such unobtrusive work, but we should remember that unobtrusiveness did not always imply artlessness. Here the bright lighting sets the mood for slapstick, while allowing us to keep the foreground and background in mind.

Perhaps the most remarkable thing about Lundin's work is the way that it increased in subtlety over the course of the decade. In *Girl Shy* Lundin is occasionally overeager to demonstrate his command of the soft style. For

Figure 4.11 Extra lighting in the foreground helps separate Jobyna Ralston from the background in *The Kid Brother* (1927).

instance, in figure 4.8, the diffusion is so strong that it blocks out most of the background. A few years later, in *The Kid Brother* (Ted Wilde and J. A. Howe, 1927), Lundin uses a different set of techniques to accomplish a similar set of functions. Following a few familiar composition conventions, he puts a little extra light on the foreground, while keeping the background in soft focus (fig. 4.11). As in *Girl Shy*, the goal is to direct our attention to Ralston, while producing an illusion of roundness. By keeping the background visible, Lundin situates the character within the narrative context, allowing us to draw parallels between the reactions of Ralston's character (Mary) and the reactions of the townspeople.

Lundin's melodramatic style also changed over the years, shifting from strong to moderate contrasts without losing the atmosphere. In figure 4.12, from *Haunted Spooks* (Alfred Goulding and Hal Roach, 1920), a frightening situation is photographed with bright highlights and dark shadows. Compare this image with figure 4.13, from *The Kid Brother*. This is still a high-contrast image, suitable for the demands of melodrama, but the shadows are no longer impenetrable. Instead, Lundin has added just enough light to maintain detail in the dark areas of the frame. Like De-Mille and Wyckoff, Lundin has decided to avoid the extremes, in order to enhance the illusion of roundness. We should resist the urge to describe this as an act of containment, as if the aggressive contrasts of the earlier film have been harnessed by the craft of classicism. Lundin fully embraced the aesthetic discourse of the ASC, and he actively sought opportunities to demonstrate his artistic skills. Given this context, it is very plausible to suppose that he viewed the more moderate style of *The Kid*

Brother as a step forward in artistic accomplishment. In the earlier film, the strong contrasts set the right mood for the genre, but they failed to provide the modeling we expect from figure-lighting, and they failed to create the sense of roundness we expect from compositional lighting. (For instance, notice the way the woman's black coat blends into the dark background.) In the later film, Lundin has unified the strategies of figure-lighting, effect-lighting, genre/scene lighting, and compositional lighting. The figure-lighting is an intensified version of the conventions for lighting men, with the actor's features modeled to the point of grotesquerie. The dark shadows set the mood for melodrama, while imitating the appearance of an abandoned interior space. Meanwhile, the shadows are just bright enough to maintain the illusion of roundness: the actor is separated from the background, and the background maintains its texture. The image from *Haunted Spooks* may be more striking at first glance,

Figure 4.12 Dark pools of shadow in *Haunted Spooks* (1920).

Figure 4.13 *The Kid Brother.* Strong modeling makes the villain's face appear grotesque.

but the image from *The Kid Brother* is much more multifunctional. As such, it does a better job fulfilling the aesthetic mandate of the ASC.

◉ ◉ ◉

In part II we will see how the normative ideals of Hollywood cinematographers changed after the transition to sound. Those ideals grew more classical, as cinematographers learned to embrace the ideals of invisibility and illusionism. However, Hollywood lighting never became a completely homogeneous style. In discourse and in practice, cinematographers continued to believe that the most advanced lighting conventions were conventions of differentiation, allowing expert cinematographers to differentiate day from night, romance from melodrama, and men from women. This insistence on the conventions of differentiation brought a new problem to the fore: the problem of conflicting conventions. Hollywood lighting grew even more complex after the transition to sound, largely because so many different cinematographers found so many different ways of solving that problem.

Part II

Classical Hollywood Lighting

Inventing the Observer

5

I n 1935 *American Cinematographer* ran a series of articles by A. Lindsley Lane, a camera operator at MGM. In an article about the camera's "omniscient eye," Lane writes:

> "All-seeingness" here means that the camera stimulates, through correct choice of subject matter and set-up, the sense within the percipient of "being at the most vital part of the experience—at the most advantageous point of perception" throughout the picture. . . .
>
> For the reason that genuine art conceals its own formulation, it may be said that a motion picture which in its showing gives self-evidence of its making is not a good picture artistically and holds the chemistry of dissolution within its own structure, drawing the audience's attention away from its story-experience purpose; is, in other words, destructive to intactness of the "illusion of occurrence," which illusion is the psychological key to a successful motion picture–percipient experience.[1]

Historians have frequently quoted this passage, largely because it ties together three of the most important components of the classical style: the commitment to storytelling, the ethic of invisibility, and the idea that the spectator is an ideal observer experiencing the world of the film through an illusion of presence.[2] Lane does more than just mention these three ideas; he also explains how they are connected. An invisible style supports the illusion of presence, allowing the spectator to pass seamlessly from one ideal viewpoint to another. This, in turn, facilitates the spectator's engagement with the unfolding narrative. Storytelling, invisibility, illusionism—here are three core concepts of the classical Hollywood cinema.

I argued in part I that the rhetoric of light did not always correspond to the ideals of classicism—at least not during the formative years of the silent period. The ASC encouraged cinematographers to think of themselves as storytellers, but it did not consistently champion the ideals of invisibility and the illusion of presence. The oft-quoted passage from A. Lindsley Lane presents a challenge to this argument, since it appears to be an unambiguous endorsement of the ideals of classicism. It is possible to meet this challenge in three different ways. The first option is to take a step back and acknowledge that the cinematographers of the silent period may have been more classical than I have been suggesting so far. Perhaps Lane is articulating a set of assumptions that were actually fairly commonplace in previous years. Indeed, it is not hard to find potential precursors to Lane, as in the comments of L. Guy Wilky. Looking to the related arts, we can note that theorists of the theater occasionally cited the illusion of presence as an ideal. For instance, in a 1918 article Arthur Hopkins argues that the theater director must become the servant of the play, employing unobtrusive techniques to create a "sustained illusion."[3] Looking farther back in time, we recall that the illusion of presence was a common theme in the discourse of painting.[4]

This is a sensible option, but I do not think we should take it very far. The ASC was the first organization of its kind. It was committed to its strategy of producing a new public identity for the cinematographer, but it did not have an obvious model for applying this strategy. Given this context, it makes sense to suppose that the ASC would have adopted an eclectic strategy during its formative years, advocating a variety of aesthetic ideals, some classical, some not so classical. Loosely defined ideals made it possible for the ASC to operate inclusively—hedging its bets, appealing to the broadest possible membership with porously defined ideals, and ensuring the organization's survival in the face of advancements both technical and artistic in nature.

The second option is to dismiss the Lane piece as an anomaly, arguing that Lane's insistence on invisibility is simply inconsistent with the aesthetic ambitions of the ASC. After all, Lane himself was an operator, not a practicing cinematographer. While this is a tempting notion, the evidence does not support it. The fact that the magazine ran a three-part series of lengthy articles by Lane is evidence that the ASC probably endorsed his ideas. More important, we can find similarly classical ideas in many other sources from the thirties. The idea that style should be invisible became a cliché of cinematographic discourse.

The third option is to propose that the rhetoric of light had changed over the years, becoming more classical after the transition to sound. I will follow this approach here, arguing the discourse and practice of Hollywood lighting became more homogeneous during the thirties, as cinematographers converged on the ideals of the classical Hollywood cinema.

One could tell this story as a tale of defeat, with the ASC surrendering its artistic ambitions to the unquestioned dominance of the classical norms, but it is better seen as a tale of a complex negotiation. To understand this shift, we must take a closer look at the array of challenges that cinematographers faced during the late twenties and early thirties. The most notable change was of course the transition to sound. Traditionally historians of the cinema had presented the transition to sound as a sudden revolution. According to this view, the tremendous success of *The Jazz Singer* forced Hollywood to make the transition overnight. Recent historians have revised the traditional views in several ways, pointing out that *The Jazz Singer* was not the runaway success of legend, that there was considerable planning prior to the transition, and that the transition itself took years. Perhaps most important, filmmakers did not see this process as a revolution; they worked hard to maintain the continuity of the Hollywood style, adapting the new technology to serve established functions.[5]

The addition of recorded sound was just one of several related changes occurring during the late twenties. Drawing on the work of scholars such as Barry Salt, David Bordwell, and Kristin Thompson, we can group the technological changes under the headings of camerawork, lighting, film stock, and lab work.[6] First, after the transition to sound, accurate synchronization required motorized cameras, as opposed to the hand-cranked models of the silent years. This minor change brought an immediate benefit to the cinematographer: he no longer had to worry about being labeled a "crank-turner." An even more important shift came during the period of multicamera shooting. In the first few years after the transition to sound, sound engineers had difficulty synchronizing the sound from different takes. The solution was to shoot a scene with multiple cameras. Multicamera shooting allowed the cinematographer to relinquish his role as a "manual laborer" altogether, instead becoming a supervisor for a team of camera operators.

While cinematographers welcomed these changes, the transition to sound introduced several aggravating challenges. Multicamera shooting required more careful consideration of lens choice, as cinematographers began to rely more heavily on telephoto lenses to provide close-ups. Sound filming also made camera movement much more difficult—initially, the cameras needed to be confined inside soundproof booths. Fortunately, most of these were temporary obstacles. The multicamera period ended in the early thirties, and inventive technicians soon solved all the major problems surrounding camera movement.

With multicamera shooting, the cinematographer had the challenge of lighting for several distinct points of view rather than just one. Again, this problem went away with the return to single-camera shooting, but a more lasting change involved the lighting equipment itself. The buzz of traditional arc lamps was too loud for sound work, and this accelerated the shift to incandescent lamps that had begun in the late twenties. The new lamps, "inkies," had distinct characteristics. Incandescent lamps typically produced a softer light than arc lamps, and cinematographers were forced to adjust accordingly.

The use of incandescent lamps facilitated another switch that was beginning before the transition: the switch from orthochromatic film stock (which was responsive primarily to the blue end of the spectrum) to panchromatic film stock (which was designed to be more responsive to the entire visible spectrum). The blue light of the arc lamps was well suited to the orthochromatic stock, but incandescents were more compatible with panchromatic. The rhetoric of realism provided a partial motivation for the switch to panchromatic stock. Orthochromatic stock had forced cinematographers to work within a narrow range of the light spectrum, but panchromatic stock carried the promise of a more "accurate" reproduction of the light of the world.

A final point concerning the changing technology is that the requirements of sound recording forced laboratories to alter their practices. Sound required more standardized lab practices, and some have suggested that this increasing standardization made it more difficult for cinematographers to experiment with unusual developing and printing techniques.[7] Even routine procedures like tinting and toning soon became obsolete. As Charles Clarke observed, "In the silent-film productions most of the theater prints were put through additional solutions and were tinted and toned. Many beautiful effects were obtained by converting the black and white image into sepia, blue or green tones by chemical agents. . . . I sometimes think it is a pity that the art of toning has been abandoned today."[8] Clarke is nostalgic for the lost beauty of the silent screen. These new restrictions on lab work would push the extravagant pictorialism of the late silent cinema toward the margins.

In addition to these changes in technology, cinematographers experienced several institutional changes. During this period the ASC began to interact more closely with other Hollywood associations. The Society of Motion Picture Engineers had been holding meetings since the teens, and some of the more technologically oriented cinematographers began to attend. The interaction with sound engineers was not always peaceful. Sound engineers had suddenly become the most important technicians on the set, and cinematographers would need to work hard to regain their previous status. Partly in response to the tensions between different branches of the industry, the Academy of Motion Picture Arts and Sciences encouraged more interaction. For instance, when studios switched from arc lighting to incandescent lighting, the Academy sponsored a meeting that allowed cinematographers and lighting technicians to study the new incandescent lights more closely.[9]

In 1928 the cinematographers formed a union with still photographers, newsreel cameramen, and special process operators. The following year the union started a publication of its own, *International Photographer*. At first the journal seemed promising, offering the possibility of a fresh voice from a union point of view. Whereas *American Cinematographer* emphasized aesthetics and technology, an early issue of *International Photographer* raised concerns about Hollywood's abusive labor practices, noting that long hours would inevitably reduce quality and efficiency.[10] Unfortunately, the magazine did not develop a distinctive voice, in part because Silas Snyder soon took over the editorship of the publication—the same Silas Snyder who had twice served as editor of *American Cinematographer* in the silent period. At a time when cinematographers and other camera operators could have used a strong union, the magazine offered lightweight articles and the occasional strained joke ("Why shouldn't an 'Inkey' give us a dark light instead of a yellow one?").[11] Meanwhile, the ASC, under the leadership of MGM's camera department head John Arnold, continued to encourage cinematographers to think of themselves as managers, rather than workers. Many of the initial union leaders struggled to find work—Alvin Wyckoff's career went into a prolonged slump, and L. Guy Wilky never shot another film after 1928. A few years later, as Mike Nielsen reports, many cinematographers failed to support a major strike in 1933, greatly weakening all of the unions in Hollywood in exchange for lucrative contracts negotiated with the aid of the ASC.[12] The early history of the union suggests that the ASC had succeeded in its goal of defining the cinematographer as a member of an elite professional class—and that its success could have some negative consequences for other film workers.

In the first chapters of this book I argued that such class identity was not just a matter of words: the rhetoric of light helped shape the visual appearance of films by encouraging cinematographers to adopt aesthetically significant lighting conventions. How would cinematographers change their rhetoric in response to the increasing institutionalization of Hollywood? This question calls for a nuanced response. On the one hand, we should expect a certain amount of homogenization, with the ASC moving closer to the ideals of other institutions, like the SMPE. The illusion of presence was a core ideal for the engineers, and the transition to sound encouraged many filmmakers to think more carefully about the illusionism of the cinema. By adopting this ideal as their own, the cinematographers of the ASC were accepting the fact that Hollywood was not in the business of selling pretty pictures. On the other hand, we should not think of this as an act of surrender, with cinematographers abandoning their ideals in the face of the powerful forces of classicism. Instead, the ASC embraced the related ideals of storytelling, invisibility, and illusionism because doing so allowed the institution to create a more compelling portrait of the cinematographer as an artist. This may seem paradoxical. Artists usually do not crave invisibility. The paradox disappears when we remember that the ASC defined storytelling in an expansive way: not just as the clear presentation of narrative information, but also as the evocation of story-specific moods, the creation of story-specific lighting effects, and the use of story-specific techniques of characterization.

To say that the ASC embraced the ideals of classicism is not to say that the ASC encouraged its cinematographers to adopt a single neutral style. Quite the contrary: as we have already seen, many of the basic lighting conventions were conventions of differentiation. After the transition to sound ASC publications explained those conventions in more detail than ever. The result was an increase in conventionalization—and therefore an increase in differentiation. Far from being forced to work in a single neutral style, the cinematographer of the thirties had more options than ever before.

When considering the ASC's relationship to other Hollywood institutions, we should not imagine the institutions as a set of concentric circles all centered on a single conception of the classical norm. It seems more useful to think of Hollywood institutions as a set of overlapping circles, as in a Venn diagram. We do not have total separation, but neither do we have perfect convergence. The ASC, the studios, the union, the Academy, and the SMPE shared a common goal—they were all committed to the production of financially successful feature-length films. Because they collaborated to make a particular product, we should expect them to have agreed about certain basic principles of construction. Still, their interests did not completely coincide, and different sections of the industry would have interpreted those principles in slightly different ways.

For instance, the ASC and the SMPE had different agendas, and different discourses to support those agendas. The writings of Joseph Dubray provide a good point of comparison. During the silent period, Dubray was a successful cinematographer and a regular contributor to *American Cinematographer*. By the thirties Dubray had left cinematography to take a job with the Bell and Howell Company. In a previously quoted passage from 1922, Dubray clearly invoked the emerging aesthetic ideals of the ASC: "The work of the cinematographer is complex. Not only good photography is required from him, but also an artistic personality, a good knowledge of dramatic construction, and of the psychological value of the graphic presentation of a dramatic or comical situation."[13] Dubray offers an early example of the mechanics-to-artists narrative, presenting the cinematographer as a man of culture.

In 1930 Dubray co-authored an article with his then-employer, A. S. Howell. Since the addition of a sound track had temporarily changed the aspect ratio, engineers were debating the virtues and drawbacks of various screen sizes. Writing in the *Journal of the Society of Motion Picture Engineers*, Dubray and Howell offer a general account of the cinema:

> The main function of motion pictures is to give a faithful reproduction of life. It is true that incidents are dramatized, that more emphasis is given to details, that outdoor scenes are selected with an eye to scenic beauty, and that interiors are always chosen, dressed and decorated in accord with the general theme of the story and the personalities of the characters which are the human elements representing what we could call an exaltation of the emotions. However the exposition of this essence of life through motion pictures demands truth of presentation and naturalness in even its most intimate details. An ideal motion picture production is one which causes the onlooker to forget his own personality and make him live with the characters of the story and in their ambient. If this psychological effect is not reached, the picture is classed as indifferent, if not entirely bad.[14]

Dubray has not changed his opinions; he has, however, changed his emphasis. In the *American Cinematographer* article, he made a passing reference to "good photography," while privileging the more artistic value of dramatic expression. In this later article, various cinematographic contributions are acknowledged, but illusionism is upheld as the supreme ideal.

When Dubray and Howell mention illusionism, they do not mean simply a painterly illusion of roundness. Rather, this is clearly a reference to the illusion of presence, in which spectators are transported to a place where they can live with the characters inside the fictional world. During this period the discourse of presence was a recurring theme in the SMPE's discourse. For instance, in a 1934 talk published the following year in the *Journal of the Society of Motion Picture Engineers*, theater architect Ben Schlanger offers a surprisingly detailed account of the illusion of presence. Discussing a new proposal for the shape of the motion picture image, he argues, "First, its shape must conform more nearly to the shape of the natural field of vision, and, second, there must appear within this shape areas of both central and peripheral vision."[15] To achieve the first goal, he proposes a large screen with approximately a 2-to-1 ratio. To achieve the second goal, he suggests that only a small central portion of this larger screen should be kept in focus. The rest of the image should be in soft focus, to suggest the blurriness of peripheral vision. This periphery would then blend into an irregular black shadow, to better disguise the supposedly distracting rectangular border of the screen. The purpose of all this was to give the spectator the illusion of seeing the fictional world directly, without the mediation of the screen.

Contrast this proposal with one that appeared in the 1930 edition of the ASC's *Cinematographic Annual*. According to cinematographer and ASC president John F. Seitz:

> The ideal screen, the screen that the near or distant future will evolve, will be a great circle, filling the entire proscenium arch of the theatre. Inscribed within this circle will be the picture, of the size, shape and proportions that best frame it. The sizes and shapes will change at will, whenever necessary. The eyes will never grow weary as these changes will furnish a pleasurable motion and variety, an interest aside from that which the drama or comedy will contain.[16]

Whereas SMPE member Schlanger proposes a model based on "natural" vision, ASC member Seitz favors an aesthetically perfect circle—perhaps the Magnascope variable screen system.[17] Far from creating an illusion of presence, this screen would allow audiences to appreciate the variations in screen size as independent attractions. We might dismiss both these proposals as irrelevant, since neither one had a long-term impact on Hollywood technology or practice. However, these different proposals do help to highlight larger differences in discourse. The SMPE and the ASC were both part of the larger institution of Hollywood cinema, but they had specific agendas of their own.

Of course, neither institution was completely homogeneous. Both allowed and even encouraged a certain amount of disagreement and friendly debate. At the same 1934 meeting that featured Schlanger's presentation, Alfred N. Goldsmith, the president of the SMPE, gave a speech on "Problems in Motion Picture Engineering." He said:

> A reasonably acceptable definition can be contrived for motion picture engineering. It is the presentation of a real or imagined happening to the audience in such approach to perfection that a satisfactory illusion of actual presence at the corresponding event is created. . . . It might be objected that exact replicas of reality will not give a desired dramatic or comic effect, and that is quite true. On another occasion, I defined the motion picture industry as "vendors of illusion and sellers of glamour." This definition need not be changed. However, the task of the engineer is to create the illusion of reality. It is for the playwright, the director, the actor, and any other artists who are involved to provide the glamour by intensifying or subduing or otherwise modifying the reality to be recorded and reproduced so that the most satisfactory audience response shall be achieved.[18]

Despite their obvious similarities, the accounts of Schlanger and Goldsmith do not overlap entirely. A point made by James Lastra can help bring out this difference. In his book on sound technology and the American cinema, Lastra has argued persuasively that engineers do not always agree whether the cinema's realism is a result of the technology's recording capabilities, or a rhetorical effect that must be carefully constructed.[19] A similar disagreement can be found here. Schlanger's illusion of presence is a constructed effect produced by the manipulation of, among other things, screen size—one of the very engineering issues that fell partly under his own jurisdiction as a theater architect. Goldsmith, by contrast, was a camera engineer, and his discourse wavers between an emphasis on the illusion of presence as a constructed effect, and the illusion of presence as a product of the camera's abilities to make exact replicas of reality. Nevertheless, both engineers do seem to agree that the illusion of presence, however it is achieved, was perhaps the engineer's most important goal—one that could organize research and practice.

The discourse of presence served to advance the SMPE's interests in part by locating its members within a particular narrative of film history—one of progress. By proposing the illusion of presence as a goal, the SMPE could tell the story of engineers constantly progressing toward that goal with each new technology. The discourse of presence gave engineers a research program, but the corresponding discourse of progress gave them a sales pitch. This is not trivial, given that one of the functions of their journal was to create a market for new inventions. In addition to serving as a means of standardizing practices, the *Journal of the Society of Motion Picture Engineers* served as a forum for engineers (such as Goldsmith) who had invented new devices—devices they hoped the industry might adopt. Engineers acknowledged that the illusion of presence was a goal that could never be achieved completely. However, the unattainability of the ideal was actually a discursive advantage to the SMPE, since this always left room for the next invention—be it widescreen, sound, deep-focus, color, or 3D. In the path toward technical progress, one can always take an additional step forward.

The ASC also had a product to sell—the artistic and technical abilities of its cinematographers. During the transition to sound, cinematographers were somewhat disturbed by the newfound importance of engineering ideals. Columbia cinematographer Joseph Walker lamented, "Photography suffered. No longer would the director ask, 'How was it for camera?' Now the all-important verdict 'OK for sound!' brought an immediate 'Print it!' "[20] The new situation was challenging, but the elite cinematographers of this proud institution had no intention of abandoning their agenda. Quite the contrary: they were determined to keep the high standards of the previous period firmly in place. In some cases cinematographers called for a return to the art of pictorialism. For instance, here is how Bert Glennon reacted to the transition to sound:

> In all the thirty-odd years through which cinematography has grown from a laboratory experiment to its present commanding position among the artistic crafts, it has never been in a more anomalous nor a more dangerous position.
>
> On the one hand, it has reached a state of great mechanical perfection, but on the other hand, its artistic growth has been arrested in the past year; and on every side we see irrefutable evidence that the art of the camera, upon

which all screen art is based, is in danger of being overlooked and becoming stagnant and buried beneath a great maze of ohms, watts, amps, and thoughtlessness.

Screen audiences want beauty. They appreciate it. They need it. Haven't you seen many audiences break into spontaneous applause at the appearance on the screen of some particularly beautiful photographic scene during the last few months? I have; and it proves that the audiences want the old cinematographic beauty of the silent pictures again.[21]

It is not hard to see that some modicum of professional jealousy motivates this display of self-righteousness. Glennon's attack is an obvious attempt to reassert the authority of cinematography in the face of the threat from engineering. The significance lies in the precise language of the attack. Glennon does not argue that cinematographers should out-engineer the engineers. Instead, he reminisces about a (mostly nonexistent) glorious golden age when artful cinematography was the main attraction.

There are several other examples of this strategy of reassertion. In 1930 and 1931 the ASC produced a new publication, the *Cinematographic Annual*. John F. Seitz's introduction to the first issue (quoted above) contains references to Titian, Pheidias, and Beethoven.[22] Victor Milner's contribution, an extended essay about the practice of lighting, bears the title "Painting with Light." Both issues of the *Annual* feature reproductions of artful photographs taken by ASC members. The message was clear: the mechanics-to-artists narrative would maintain its place as the founding story of the ASC and continue to be a relevant fable in an era marked by technical advancement.

Nevertheless, the discourse was changing in subtle ways. The ASC did not abandon the mechanics-to-artists narrative, but pictorial beauty was no longer an essential ideal. Instead, the institution began to place additional emphasis on aesthetic ideals that were more compatible with classical notions like invisibility and the illusion of presence. For instance, in a 1932 article on diffusion, Paramount cinematographer Charles Lang contended:

It must also be remembered that cinematography, as used in the production of photoplays, is essentially a dramatic, narrative Art: accordingly, it must be at once consistent and unobtrusive. . . . As some great actor (I think it was George Arliss) once remarked, the secret of art is not being natural, but being unnatural—without getting caught at it. The same thing is doubly true of dramatic cinematography: its greatest secret lies in utilizing its manifold artistic and mechanical tricks to direct the emotional response of the beholder, without giving the slightest suggestion of their employment.[23]

Unlike his peers in the silent period, Lang elevates the principle of invisibility to the status of a mandate: cinematography "must" be unobtrusive.

Lang was one of the greatest cinematographers in Hollywood—a man who would soon win the Academy Award for his brilliant work on *A Farewell to Arms*. Why on earth would this man want his style to be invisible? This is a surprisingly difficult question. The simplest way to answer it would be to shift the burden to the larger Hollywood norms. Invisibility was not just a norm for cinematographers—it was upheld as an ideal throughout Hollywood, by directors, composers, editors, set designers, and more. Suppose Charles Lang secretly wanted to make abstract films. Assuming he wanted Paramount to continue giving him paychecks, he would have kept this wish to himself. Perhaps invisibility was a norm for cinematographers simply because it was a governing norm for any Hollywood practitioner.

This explanation is not without merit. The concept of "norm" should be something more than a statistical construct. To say that the lighting conventions were normative is to say that Hollywood practitioners were supposed to follow them whether or not they agreed with the principles. Given that invisibility was embraced throughout Hollywood, we should expect cinematographers to embrace it, too.

Nevertheless, this explanation is unsatisfying for two reasons. First, it implies that Hollywood's basic norms were created first and cinematographic norms followed. However, cinematographic norms were a constitutive component of Hollywood norms. If cinematographers had mounted a sustained attack on invisibility, they might not have overthrown the Hollywood norm, but they could have at least shifted its center. Instead, cinematographers supported the aesthetic of invisibility, arguing that it should be installed as an industry-wide ideal.

This brings us to the second problem with the above explanation. Most norms were not imposed by fiat. Practitioners accepted them because they found the reasons that justified them to be convincing. The purpose of cinematographic discourse was not to strong-arm cinematographers into following a set of ironclad rules, but rather to list a set of ideals and to give cinematographers good reasons for following them. While it is certainly possible that a prominent cinematographer like Lang secretly longed to make an abstract film, it seems much more likely that the vast majority of cinematographers (including Lang) accepted Hollywood's filmmaking principles in good faith. They embraced invisibility because they thought it was a positive aesthetic value.

An episode from the transition period brings out these two qualities of cinematographic norms. After the move to sound, many ambitious directors experimented with extravagant camera moves, in an effort to recapture the mobile style of late silent cinema. We might think of flamboyant camera movement as the ultimate rejection of the illusion of presence, since it tends to remind the spectator of the mechanics of camerawork. However, the situation was probably more complicated than that. For instance, the widely admired F. W. Murnau had once justified his late-silent-period movements by claiming that they were designed to follow the hypothetical movements of person observing the action:

> To me the camera represents the eye of a person, through whose mind one is watching the events on the screen. . . . It must whirl and peep and move from place to place as swiftly as thought itself, when it is necessary to exaggerate for the audience the idea or emotion that is uppermost in the mind of the character. I think the films of the future will use more and more of these "camera angles," or as I prefer to call them these "dramatic angles." They help to photograph thought.[24]

This is clearly not a perfectly pure example of classical discourse. The passage shows the influence of European trends and theories, such as the Expressionist desire to exaggerate emotions, and the Impressionist desire to photograph thought. Still, it is striking that Murnau appeals to the idea of the ideal observer to explain such a flamboyant artistic flourish. This suggests that the concept may have carried different connotations at the time. Far from being an obvious classical norm, the idea of an invisible observer was seen as daring, experimental, perhaps even futuristic. Hollywood directors may have been trying to recapture the same mixture of classical and nonclassical connotations when they began using elaborate camera movements again soon after the transition to sound.

We might expect cinematographers to have welcomed the chance to flaunt their virtuosic skills, but instead cinematographers criticized the directors for using style in such a flamboyant way. One article from *American Cinematographer* offered the opinions of various different cinematographers. James Wong Howe's response was particularly interesting. Howe was himself a master of camera movement, having designed such famous examples as the introductory shot from *Transatlantic* (William K.

Howard, 1931). In this article, Howe warns against unusual camera moves, pointing out that the camera represents the eye of the spectator. Other cinematographers offer similar complaints, connecting the rhetoric of illusionism to the rhetoric of the invisible style. Tony Gaudio accuses directors of "showing off" to the studio heads. Victor Milner argues that style should be "imperceptible."[25] Whereas Murnau's position teetered on the edges of the classical style, it appears that the cinematographers have moved to the center, rejecting flamboyant displays of style in favor of an illusionistic aesthetic.

This does not mean that cinematographers had simply submitted to the classical norm. Quite the contrary: they were using the rhetoric of the classical norms to advance their interests in a dispute with a more powerful group that appeared to be rejecting those norms—namely, directors. Cinematographers had an ulterior motive for this campaign: they defined the art of cinematography as an art of lighting, and elaborate camera moves made the art of lighting much more difficult to practice. Significantly, the cinematographers' arguments were not simply arguments about what not to do. They gave as much rhetorical attention to arguments about what style can and should do: namely, support the story by directing attention, by denoting time and space, and by enhancing the narrative's moods.

Part of the confusion stems from the fact that cinematographers are said to have "subordinated" style to story. This phrase can be very misleading. For many cinematographers, the art of Hollywood cinematography is not *subordinated* to storytelling; the art of Hollywood cinematography simply *is* an art of storytelling. As William Daniels once said, "We try to tell the story with light as the director tries to tell it with his action."[26] Style functions like an arrow, directing the audience's attention to the most important story point. Typically, the arrow's function mandates a certain structural feature—namely, that the arrow should point to something other than itself. What good is an arrow that points to itself? Hollywood cinematographers seem to have taken on this logic when they assumed that cinematographic style should not call attention to itself. Given that one of style's functions is to direct attention toward the fictional world, a reflexive style would be counterproductive.

For Charles Lang, style is not just a way to guide the audience's attention—it is also a way to guide the audience's emotions. The true art of cinematography is an art of mood and composition, using carefully selected strategies to help tell the story. Lang does not write with the tone of a humble craftsman, willingly subordinating his style to the needs of the story. For Lang, it is a matter not of subordination but of *coordination*, of finding a perfect match between story and style. If this is classicism, it is classicism with an aesthetic emphasis specific to the institutional context of the ASC. Like Walter Lundin before him, Lang does not define storytelling as simple narration—as the manipulation of the spectator's comprehension of the linear causal chain. He defines storytelling in a more expansive manner, with a particular emphasis on the use of style to direct the spectator's emotional response. In this way he is actually extending the discursive strategy of Seitz, Lundin, and Wilky. Cinematography is the art of finding the right mood for the story.

The right-mood-for-the-story theory would continue to play a crucial role in the cinematographic discourse of the thirties. We cannot understand that discourse without understanding the strong emphasis it placed on the value of expressive lighting. By invoking the concept of mood, cinematographers were self-consciously reasserting an ideal that had been established during the silent period. In an article on "Lighting" in the 1931 edition of *The Cinematography Annual*, James Wong Howe writes:

> Methods of lighting a motion picture set have changed tremendously since the day, long ago, when Cecil B. DeMille wrote that lighting to a motion picture was like music to an opera; but the importance of skillful lighting has not changed, save to increase. With the early films, lighting merely meant getting enough light upon the actors to permit photography; today it means laying a visual, emotional foundation upon which the director and his players must build. In other words, lighting has changed from a purely physical problem to an artistic, or dramatic one.[27]

Howe had worked with DeMille in the silent period, when DeMille was experimenting with the techniques of "Rembrandt lighting." Here Howe is arguing that DeMille's theatrical ideas remain just as relevant as ever. Lighting should perform a dramatic function; that is, it should enhance the emotional development of the story. This concept is not expressionistic, but it is expressive, in the sense that it involves the use of light to evoke an emotional response. The production of emotion is a long-standing artistic ideal. For Howe, achieving this ideal is the culmination of the mechanics-to-artists narrative.

During this period *American Cinematographer* printed several articles that could be classified as practical film theory, with a variety of writers attempting to understand and explain the role of expression in the art of cinematic lighting. When we take a close look at this theory of expression, we see that it was surprisingly compatible with classical ideals like the illusion of presence and the ethic of invisibility. Cinematographers were not psychologists, and they did not offer a detailed theory of the relationship between light and mood. Instead, they explained the idea of mood by appealing to the concept of association. For example, in 1934 *American Cinematographer* ran a series of articles by camera operator L. Owens Huggins. One article was entitled "The 'Language' of Tone." Huggins writes:

> With a high key, we can express happiness—gaiety—joy—delicacy, and freshness. . . .
>
> An extremely low key will suggest restraint—severity—sombreness; and sometimes grief and sordidness.
>
> A medium key with a predominance of greys, but lacking any bright highlights or extreme shadows, obviously suggests fogginess; hence, mystery and vagueness.
>
> Its opposite, an equally limited tonal scale in which the intermediate tones are suppressed, leaving only the extremes of highlight and shadow, suggests the bizarre—often the supernatural or horrible—always the unusual.[28]

Huggins's model is a simple one: any cinematic device, such as a color, line, or tone, can produce fairly predictable associations in the mind of the spectator. The device then evokes whatever emotion is typically produced by that association. Huggins is not sketching a particularly deep theory of psychology, but his ideas could provide a useful heuristic for cinematographers seeking to create an expressive style. Although Huggins himself was not a cinematographer, we can find this same associationist logic underlying many of the basic genre conventions, as in the "soft" style of the romance, the "hard" style of the melodrama, the "brilliant" style of the comedy, or the "somber" style of the drama. Decades later, Charles Clarke, a former cinematographer at Fox, wrote a textbook on the practice of cinematography. In his chapter on lighting, he appealed to the same associationist theory, writing, "An important psychological factor is also introduced, for we all associate bright, clear light with happiness and well being. With dull, heavily diffused light, we associate gloom, storm, and impending trouble. This psychological factor is also a basic law in the art of lighting."[29] Psychologists might raise doubts about the validity of this law, but it is clear that Clarke had no such doubts.

Once we suppose that there is an associationist logic underneath the discourse of expression, we see that expressive ideals were quite compatible with most of the other aesthetic ideals that shaped Hollywood lighting. As we saw in chapter 3, the conventions of effect-lighting were often justified by a rhetoric of realism. It is often assumed that realism and expressivity are incompatible goals, but the associationist theory does not mandate such an assumption. According to this theory, darkness can produce an emotional response in the spectator by suggesting certain associations. This psychological process will occur whether the darkness is motivated or not. Given that fact, there is nothing preventing the filmmakers from motivating the darkness by having a character turn off a lamp or close the window. As Barry Salt has suggested, light can be expressive, evoking an emotion, without the distortions characteristic of Expressionism. By combining the genre/scene conventions with the effect-lighting conventions, the cinematographer can better fulfill the ASC's mandate, accomplishing multiple aesthetic goals at the same time.

Even more important, expressive techniques are fully compatible with the techniques of storytelling. The story is more than just a causal chain; it is the generator of emotionally charged situations. The cinematographer's job is to align the progression of the mood lighting with the emotional progression of the story. Some writers took this idea of coordination even farther, almost to the point of suggesting that the spectator is watching two films at once. The most explicit statement of this idea came from Slavko Vorkapich, whose essays appeared fairly regularly in ASC publications. In an essay called "Cinematics," he writes:

> A perfect motion picture would be comparable to a symphony. It would have a definite rhythmical pattern, each of its movements would correspond to the mood of the sequence and each individual phrase (scene) would be an organic part of the whole. And like a symphony it would be interesting at every moment of its development regardless of the meaning or story it has to convey. In other words: a motion picture should be visually interesting even if we entered the theatre in the middle of the performance; we should be visually entertained even if we did not know the beginning of the story.[30]

Vorkapich might say that the spectator is watching two films simultaneously: a fiction film in which the events of the narrative are seen in a three-dimensional world; and an abstract film in which two-dimensional graphic qualities produce a visual symphony. The task of the cinematographer is to coordinate these two films, by making the mood of the graphic pattern match the mood of the narrative.[31]

We might reject Vorkapich as an avant-garde aesthete, but his ideas inform essays written by much more conventional figures. Victor Milner was one of the most frequent contributors to *American Cinematographer*, and his articles routinely appeal to the discourse of expression. In one article he writes that the cinematographer "can attune the visual mood of the picture to the dramatic mood of the story."[32] This suggests that the picture and the story create moods independently of one another in theory, although they can (and should) be coordinated in practice. In a later article on "Creating Moods with Light," Milner again gives hints of Vorkapich. He writes:

> Before the advent of sound and dialog, 90% of the responsibility of securing the desired emotional effect in such a scene lay in the hands of the Cinematographer. Today, even with the great advantages of speech, vocal inflection, music, and sound-effects, it still rests with the Cinematographer whether the scene shall be merely a well-acted scene, or a gripping emotional experience. The true test of Cinematography is the emotional and dramatic effect it would convey if viewed without the assistance of the sound-track.[33]

The cinematographer certainly does produce an emotional effect with the aid of a fully constructed narrative world, but a good cinematographer can also produce such an effect without it. Like Vorkapich, Milner argues that the effects of cinematic imagery are logically distinct from the effects of the narrative. To say that Milner subordinates his style to story is to underestimate the self-confidence of his discourse: Milner happily takes credit for 90 percent of the silent film's emotional effectiveness, and he believes that a sound film would retain much of its effectiveness with the sound turned off.

I am not suggesting that Milner was a closet modernist, secretly rebelling against the classical style. Milner's discourse was consistent with the general outlines of classical discourse, even as he gave specific details an idiosyncratic interpretation that was more consistent with the institutionally constructed identity of the elite Hollywood cinematographer. In spite of his self-confident rhetoric, Milner does not argue that style should be "visible." Expressive lighting works by creating a "subconscious, emotional receptiveness." There is no need to foreground style when it can do its work in more unobtrusive ways. The discourse of expressivity and the discourse of invisibility support each other.[34]

This is the context that can allow us to understand a theorist like A. Lindsley Lane, quoted at the beginning of the chapter. Lane is encouraging cinematographers to adopt classical ideals, without asking them to abandon their ambitions of artistry. Many cinematographers were attempting to accomplish a similar balancing act, combining the aesthetic idealism of the ASC with the new demands of an increasingly institutionalized Hollywood industry. We do not have a set of concentric circles here, with all the members of Hollywood converging on a single norm of unobtrusive narration. Instead, we have overlapping circles, with most institutions sharing the ideal of unobtrusive storytelling, but with each institution interpreting that ideal in its own way.

◎ ◎ ◎

It may seem as though I am proposing merely some minor modifications to the classical model, but these minor modifications could have major consequences for the ways that we look at Hollywood lighting. Many historians of Hollywood style are fascinated by filmmakers who craft eye-catching images of pictorial beauty—filmmakers like Josef von Sternberg and his favorite cinematographer, Lee Garmes. Von Sternberg is interesting precisely because his flamboyant pictorialism goes beyond the outer limits of the classical style. The problem is that an emphasis on the artfulness of the von Sternberg style inevitably relies on a contrast with the supposed artlessness of the typical Hollywood style. It is too easy to construct the standard Hollywood style as a neutral norm, only to shift our attention to obvious exceptions. It is much more challenging to understand how the ASC managed to embrace the ideal of unobtrusive storytelling and affirm the artistry of Hollywood cinematography at the same time. This is an intriguing paradox. To fully understand this paradox, we must now shift our attention away from the eye-catching ideals of pictorial beauty and toward a set of openly aesthetic ideals that were more compatible with the notion that the art of cinematography is an art of storytelling—ideals like characterization, realism-of-detail, compositional clarity, and, especially, mood.

6 Conventions and Functions

Rather than argue that lighting should serve a single function, the American Society of Cinematographers consistently argued that lighting could serve several different functions. We might divide those functions into four general categories. First, lighting can help tell the story by directing the spectator's attention, setting an appropriate mood, denoting time and space, and enhancing characterization. Second, lighting can serve the ideal of realism—an ideal that cinematographers defined in various ways, from the realism of detail to the realism of illusionism. Third, lighting can improve pictorial quality, by adding beauty or foregrounding innovation. Fourth, lighting can glamorize the star—a function that would occasionally run the risk of undermining the other three. Clearly, the mechanics-to-artists narrative allowed for several different paths to artistry.

As I argued in the introduction, there is no one-to-one correlation between functions and conventions. The most commonplace conventions were widely used because they accomplished a variety of functions at the same time. Scholars such as Barry Salt have done an excellent job explaining how the style of Hollywood lighting changed in subtle ways as a result of the technological advances that occurred in the late twenties and early thirties. Without meaning to diminish the importance of those technological changes, I will show here that patterns of continuity were equally important. Just as the ASC reasserted its aesthetic ideals in response to the challenges of the transition, the cinematographers of Hollywood reaffirmed their commitment to the four sets of conventions they had established during the silent period, refining them by making the rules even more complex and multifunctional.

Figure-Lighting

As we saw in chapter 2, three-point lighting was the basis of Hollywood's figure-lighting conventions. During the multicamera period, even this basic technique was difficult to practice. With multiple cameras, a cinematographer would often need to shoot a close-up and a wide shot at the same time, from different angles. The problem is that the same lighting set-up looks different when viewed from different places on set. For instance, the glowing effect of a backlight is specific to a certain point of view. When seen from another angle, the glow is much less pronounced. In addition, the use of multiple cameras restricts the available options for light placement. A light that would be just out of frame in a close-up might be in the middle of the frame for a wide shot. To make matters even more complicated, cinematographers had to work with sound engineers to address new problems like microphone shadows and buzzing light fixtures. In general the transitional period can be seen as a setback for Hollywood's figure-lighting conventions. As Charles Clarke noted, "This produced a flat, general lighting with a minimum of pictorial interest. For a time, all the fine quality of lighting we had developed in the silent films went down the drain."[1] Still, cinematographers tried to use the three-point technique whenever possible, occasionally shooting glamorous silent close-ups that could be inserted as cutaways during dialogue scenes.

After the return to single-camera filming, three-point lighting was once again installed as the default convention. Some cinematographers simply used this technique for every character, regardless of other considerations. However, the ASC continued to advocate a more "artistic" approach: lighting for character. As in the silent period, character was defined in gendered terms. For instance, in one 1930 article, William Stull writes:

> Close-shots of people can be not only records of their physical appearance, but artistic portrayals of their characters, as well. Men, for instance, are best photographed with rather hard lightings, and in sharp focus. This lends a virile, masculine quality to the scene. Women, on the other hand, are often better shown with softer, flatter lightings—especially well-balanced back-lights—and in soft-focus. This accentuates the feminine gentleness.[2]

Although Stull goes on to qualify this advice by insisting that these rules are only generalizations, the basic advice is strikingly similar to the ideas that cinematographers had inherited from portrait photographers during the silent period. In particular, Stull links together a triangle of concepts: artistry, character, and gender. Hollywood figure-lighting is artistic because the cinematographer uses techniques to portray character, which is understood as a basic contrast between masculine hardness with feminine softness.

The same ideology permeates a 1932 article on "Shadows," which recommends careful control of contrast:

> When photographing women they should be done so beautifully. The lighting should be in a high key and aim to express femininity. The tonal range between the highlight and the shadow should never be very great.
>
> The lighting for men on the other hand should express rugged virility. The tonal contrast should be much longer than that employed for women. In fact it should be more or less contrasty without being violent.[3]

This article appeals to the ideology of character to justify a rule concerning the use of fill light. A woman's face would feature "gentle" gradations, while a man's face would express virility with stronger contrasts. In these passages we have a set of recommendations regarding direction of light, quality of light, contrast, and lens use. Figures 6.1 and 6.2, from *Bullets or Ballots* (William Keighley, 1936), demonstrate the familiar conventions governing the placement of key-lights. Cinematographer Hal Mohr gives Joan Blondell a frontal key, flattening her features. He gives Edward G. Robinson more modeling by placing his key in a cross-frontal position.

There is evidence that Warner's head of production, Hal Wallis, was aware of this convention. During the filming of Raoul Walsh's *The Roar-*

Figure 6.1 The front-cross key models the face of Edward G. Robinson in *Bullets or Ballots* (1936).

Figure 6.2 In the same scene, a more frontal key for Joan Blondell.

ing Twenties (1939), Wallis wrote a revealing memo regarding cinematographer Ernest Haller. Haller was Bette Davis's favorite D.P., but he was assigned to shoot the Cagney gangster picture, and Wallis was not pleased with his work: "Let him give some character lighting to Cagney in the closeups, instead of making him look so beautiful. See that it is done in sketchy lighting, in shadows, etc."[4] Perhaps because executives were aware of the convention and ready to enforce it, Warner Bros. cinematographers seem to be particularly faithful to this rule regarding key placement. It appears in a range of Warner Bros. films, from the minor romance *The Rich Are Always with Us* (1932, shot by Ernest Haller) to the major adventure film *The Charge of the Light Brigade* (1936, shot by Sol Polito). The placement convention was operative at other studios as well, structuring the cinematography for films like Universal's *The Invisible Man* (1933, shot by Arthur Edeson) and Paramount's *Midnight* (1939, shot by Charles Lang).

The placement convention could also apply to two-shots. Charles Clarke explains the convention clearly:

> We frequently arrange the key light from the side that the leading lady will face, so that she will have the benefit of the more front light (for she is thus facing the three-quarter key), while the male player will be facing away from it. He would look well with that kick-light technique, which would also be a backlight for her. We would project a fill light on him so that he would not be in complete shadow.[5]

Figure 6.3 shows an example, from *Anna Karenina* (Clarence Brown, 1935). The key-light is to the left of the camera, smoothing the features of

Figure 6.3 The key-light is to the left of the camera, favoring Greta Garbo in *Anna Karenina* (1935).

Greta Garbo's face while leaving Reginald Denny's face in shadow. Cinematographer William Daniels compensates for the shadow on Denny's face by adding plenty of fill, and using the backlight to bring out the details of his eyes, nose, and mouth. This backlight also serves to add shine to Garbo's hair. Daniels uses a similar technique throughout the film. When the shot features two men or two women, Daniels points the light at the bigger star.

In silent films like *The Mysterious Lady* (1928), Daniels had mastered the techniques of the soft style, but he is using much less diffusion here. Indeed, almost all Hollywood cinematographers used lens diffusion less and less as the decade progressed. This is partly because the transition to sound increased the interest in realism, but it may also be a result of the changing attitudes concerning the essential "softness" of the feminine character. The soft-focus shots of Lillian Gish in *Way Down East* seemed as old-fashioned as the film's small-town protagonist, Anna. Still, the hard-soft distinction did not disappear completely. A cinematographer might choose to shoot a woman in sharp focus, while employing extra diffusion on the lamps. Or a cinematographer could create a "softer" style by controlling the contrast with fill lighting. In DeMille's trash masterpiece *The Sign of the Cross* (1932, shot by Karl Struss), Elissa Landi receives more fill light than Fredric March. The result is a softer style for her, with gentle gradations, while March's face is etched with strong contrasts. An even more rigorous example of the control of contrast might be Leon Shamroy's work on Fritz Lang's *You Only Live Once* (1938). No matter where Sylvia Sidney goes, she always seems to look softer than Henry Fonda.

In exteriors, it was difficult to control the lighting with this degree of precision, since no cinematographer could ask his gaffer to move the sun.

In figure 6.4, from Henry Hathaway's *Peter Ibbetson* (1935, shot by Charles Lang), the sun is overhead, producing an unflattering top-light on the two players, with a fair amount of bounce-light from below. Such a lighting scheme would be totally unacceptable for a close-up, and Lang softens the sunlight considerably when the film cuts in for a romantic two-shot, probably by stretching a large diffusing silk over the actors' heads (fig. 6.5). The actors are positioned in such a way that the lighting recreates the standard two-shot strategy for a male/female pairing, with the key-light favoring the woman's face. Another option was to use the sun as a backlight, while using a reflector or an arc to produce the key. In a 1932 article, Charles Clarke describes the wide variety of reflectors available to the cinematographer, such as powdered aluminum, white paint, metallic gold powder, and tinned metal sheets. Significantly, he explains that some reflectors are softer than others, explicitly urging cinematographers to use the softer reflectors for close-ups of women.[6]

Figure 6.4 *Peter Ibbetson* (1935): In a medium-long-shot, the sun casts a hard light from above.

Figure 6.5 The lighting is softer in the closer view.

These examples suggest that cinematographers of the thirties were fully committed to the strategy of differentiation, even if the Lillian Gish ideal was falling out of favor. However, the ideal of glamour worked against the trend toward differentiation. While the desire to glamorize a star could encourage cinematographers to emphasize certain individual features (such as Garbo's eyelashes or Dietrich's cheekbones), glamour was more likely to be a homogenizing force, asking cinematographers to paint everyone with the same beautiful brush. In fact, Hollywood cinematographers were occasionally criticized for lighting men in an overly feminine manner. In a Russian book on filmmaking, there is an unusual passage in which author Vladimir Nilsen praises Soviet cinematographers by noting that they routinely give their male leads more "virile" treatment than do their peers in Hollywood, who are accused of weakening the male leads by bathing them in "feminine" glamour.[7] While this passage may reveal more about the values of Nilsen than it does about the comparative merits of Soviet and Hollywood cinematography, it is true that very few of Hollywood's star actors were receiving the rugged treatment that Joseph August was giving to William S. Hart back in 1920.

On this point studio styles are particularly significant. MGM was the most glamorous studio in Hollywood. Because most of the MGM stars receive the unrestrained glamour treatment, there is often surprisingly little difference between the lighting strategies for male and female stars. For instance, in the 1936 version of *Romeo and Juliet*, Leslie Howard's Romeo is not lit very differently from Norma Shearer's Juliet. Both typically receive skin-smoothing frontal keys, generous amounts of fill, and halo-producing backlights. (This glamorous style also helps to hide the fact that Howard and Shearer are much too old to be playing the star-crossed teen lovers!)

By contrast, Warner Bros. was known for its tough-guy stars, and it is not surprising that this studio placed more emphasis on character lighting for men. These differences were actually reinforced by differences in the studios' respective laboratory practices. As Barry Salt has shown, it was customary for MGM cinematographers to deliberately overexpose their images. The laboratory would then underdevelop the stock, producing a low-contrast image with gentle shades of gray. Meanwhile, Warner Bros. cinematographers would practice intentional underexposure. When the laboratory compensated by overdeveloping the stock, the result would be a high-contrast image with darker shades of inky black. The reasoning was partly economic: MGM could afford to expend electricity on extra light, while the financially strapped Warner Bros. could not. However, economics was not the only reason for the strategy. The softer style was more appropriate for a studio that specialized in romantic glamour pics, while hard contrast was suitable for a studio with a stable of tough-guy stars.[8]

Still, glamour was a mandate at all the studios, from Paramount to Columbia. Many cinematographers were surprisingly hostile to this mandate, believing that it prevented them from fulfilling an equally important mandate: storytelling. Indeed, this tension between the glamour function and the storytelling function was one of the defining conflicts of Hollywood cinematography during the classical period. We will take a closer look at this conflict in chapter 7, but for now the important point is to note that the ASC continued to discuss the problem of figure-lighting with the rhetoric of artistry. The cinematographer was not simply a servant of the stars—he was an etcher of character.

Effect-Lighting

The conventions of effect-lighting developed through a gradual process of accumulation, with cinematographers constantly searching for new ways to represent moonlight, sunlight, and candlelight. Technology facilitated this expansion. The first artificial lights were floodlights, which cast a bright light uniformly over a wide area. Soon cinematographers began to use spotlights, which are capable of projecting a narrower beam. In the late twenties, the expansion of effect-lighting was aided by the switch from arc lights to incandescents, since the notoriously dangerous arc lights required more careful attention. At several points in the following decade, Kodak introduced faster film stocks, supporting the expansion of effect-lighting by allowing the use of smaller spotlights to produce a visible exposure. Eventually, by the late thirties, the lights got so small that a clever cinematographer could hide a tiny spotlight in the middle of the shot, to suggest the light of a candle or lamp. When engineers designed arc lights that were easier to use in sound production, cinematographers could choose between the sharp shadows of the arcs and the slightly softer shadows of the inkies. Meanwhile, grip equipment was becoming more sophisticated, allowing cinematographers to produce precisely controlled shadows with the aid of some carefully placed nets and flags.

We tend to think of lighting effects as unusual tricks that appear in minor genres like the horror film. However, effects were thoroughly commonplace throughout the studio period. For instance, consider a 1942 article by cinematographer Phil Tannura. He writes:

> Just to get the record straight, my impression of the meaning of the term "effect-lightings" would run something like this: it's any type of lighting which attempts to reproduce the effect of the illumination you'd actually see in any particular room or place under the conditions of the story, as apart from the smooth, overall illumination of a conventional lighting.
>
> That, you may say, covers a lot of territory—but so do effect-lightings! They can range from the most extreme and obvious instances like night-effect interiors in a mystery-story, where the room is in total darkness except for a beam of moonlight splashing through a window, to the rather less obvious effect-lightings which might reproduce, say, the effect you'd normally get in your living-room at night, with the illumination coming from one shaded lamp on the table and a reading-lamp by your pet armchair.[9]

Tannura goes on to explain how an amateur might accomplish some of the most typical professional tricks. The fact that this article appeared in the amateur section of *AC* suggests that Tannura is not proposing anything new. He is explaining a firmly established set of techniques—techniques that the average cinematographer would take for granted. In this sense, Tannura does not disclose a closely guarded magician's trick; rather, he imparts some commonplace slight-of-hand for the everyman to use while amusing parlor guests.

This article was one of several articles on effect-lighting to appear in the pages of *American Cinematographer* during the late thirties and early forties. Tony Gaudio wrote a particularly informative article in 1937, discussing the fact that the new baby spotlights were allowing for an increase in precision. According to Gaudio:

> There is a natural focal highlight in every scene. Almost always, this coincides with the center of interest of that scene. The lighting should radiate from this natural focal highlight. There may be, and almost always are secondary principal highlights, but they should be distributed in pleasing relationship to this main center of interest and light.
>
> Creating motion picture lightings from this viewpoint permits—even compels—the use of more natural lighting effects. For instance, suppose I am seated in a room at a desk. A desk light creates a strong focal highlight where I sit. Farther down the room is a window through which the light from outside creates a secondary highlight area. If I rise from the desk and walk to the win-

dow, you would see me pass through the shaded area and then enter the secondary highlight. . . .

Lit as I have been trying to light my scenes this past year, the shadow areas would be carefully lit with spotlight beams so that the illumination gradually fell off as I left the main highlight area and then increased again as I approached the second highlight area.[10]

Although Gaudio is celebrating the new technology, his argument is not entirely new. Gaudio himself had written an article calling for natural light effects way back in 1917.[11] In this article from 1937, Gaudio demonstrates that effect-lighting was still associated with rhetoric familiar from the theater. Effect-lighting is an ideal strategy because it combines multiple functions. First, effect-lighting creates zones of interest, which can be used to direct the spectator's attention to the unfolding narrative. Second, by capturing the gradual fall-off of light between the two zones, Gaudio creates a more pleasing picture. A third function is an increase in naturalness. Gaudio describes his technique with the phrase "lighting as in life," assuming that a natural lighting effect is inherently superior to an artificial one. Most important, Gaudio ties these functions together with a rhetoric of progress. Just as the discourse of the theater claimed that modern lighting was the first to become truly multifunctional, Gaudio argues that technological progress and artistic progress work hand in hand.

A shot from *The Amazing Dr. Clitterhouse* (1938, directed by Anatole Litvak) demonstrates how Gaudio applied these techniques. Throughout the shot, Gaudio uses effect-lighting to manipulate our attention while remaining within the bounds of plausibility. In the beginning of the shot, Dr. Clitterhouse (Edward G. Robinson) is watching a singer perform at a society party (fig. 6.6). Our attention is on the singer. A previous shot has already established that there are several lights in the parlor, and it is plausible that this space is brightly lit—though Gaudio has cheated a bit by putting extra light on the singer. The doctor then turns and walks to the right. He will eventually arrive at a telephone sitting by a lamp. This means that the middle portion of the walk should be somewhat dark, as he is located between two light sources. Appropriately enough, Gaudio lights the doctor from two directions (fig. 6.7). The doctor's back is lit from the left, suggesting the light from the parlor. The doctor's face is lit from the right, suggesting that he is moving toward another light source. We do not yet see the lamp, but its presence is suggested by the fact that a nearby object casts a noticeable shadow on the wall. Soon the doctor arrives at the telephone, where he stands to the right of the lamp. Gaudio's key-light comes from the left of the camera, simulating the effect of the lamp. Now the doctor is brightly lit (fig. 6.8). This befits his status as the most important part of the scene, and it is also plausibly motivated by the fact that he is standing so close to the lamp. In this shot the lighting effect is subtle and unobtrusive, but Gaudio would insist that it is still artful. As he notes in the final sentence of his article, "The results on the screen are both more artistic and more natural."[12] According to the rhetoric of effect-lighting, these two ideals are mutually supporting.

In a 1941 article celebrating the new technology, Arthur Miller draws on the same set of ideals. The article's title is "Putting Naturalness Into Modern Interior Lightings." Miller writes:

> And within the past eighteen months an even smaller lamp—the tiny 150-Watt "Dinky Inky"—has been developed, and proved itself invaluable. Before the days of fast film, such a lamp would have been too absurdly small to have any practical value. Today, it has become the fine brush by which we can at last paint our precision light-effects with the small, delicate brush strokes we have so long needed.[13]

Miller explains effect-lighting by drawing on several familiar rhetorical themes—the rhetoric of technological progress, the rhetoric of naturalness, and the rhetoric of painting. The tools of effect-lighting allow the cinematographer to paint with light. Miller illustrates his article with examples from two of his own films, *The Mark of Zorro* and *Brigham Young: Frontiersman* (both 1940).

Notice that neither Gaudio nor Miller claims that his work is artistic simply because it is innovative. They do not associate artistry with the continual production of novelty. In fact, such a move would have been antithetical to the purposes of the ASC. The ASC's goal was to encourage the advancement of the art as a whole, and it did so by encouraging cinematographers to share their most useful techniques. That is precisely what Gaudio and Miller are trying to do. By the late thirties, some light-effects had become so conventional that we can expect to find them in almost any Hollywood film. Perhaps the most commonplace effect was the technique of shining a powerful light through a window in order to imitate sunlight. The effect, appearing in films as diverse as *The Mummy* (Karl Freund, 1932) and *Bachelor Mother* (Garson Kanin, 1939), quietly performs a number of functions: denoting the time of day, adding realistic detail, keeping our attention in the center of the frame, and providing

Figure 6.6 In *The Amazing Dr. Clitterhouse* (1938), Edward G. Robinson turns to the right . . .

Figure 6.7 . . . walks through a dark area between light sources . . .

Figure 6.8 . . . and picks up a telephone next to a lamp.

a dash of pictorial pleasure. The window-pattern effect was so recogniz-able that a cinematographer might suggest sunlight streaming through a window without even bothering to show the window itself. The effect of moonlight streaming through a window was similar, but it usually called for more careful management of the film's contrasts. A cinematographer might employ a comparable level of brightness for the actual moonlight effect, but reduce the light level of the overall room to create more contrast between the portions of the wall that are illuminated by the moonlight and those that are not.

Another routine effect was the that of an indoor light source, such as a table lamp. When it comes to effect-lighting, screwing in a light bulb is a lot harder than it looks. An ordinary light bulb will be too dim, given the high light levels that filming requires. Cinematographers typically used photo-flood light bulbs, which give off a brighter amount of light for a shorter period of time. The lights cannot be too bright or they will overexpose. Choice of lampshade is also important: it must prevent lens flare while tempering the effect to reach the proper light level. Setting up the table lamp is only part of the problem. No matter how bright the table lamp may be, it is usually not bright enough to light the actors' faces. To complete the effect, the cinematographer must arrange the key-light so that it appears to come from the general direction of the table lamp.[14]

The window pattern and the table lamp were probably the two most common lighting-effects in Hollywood film. Of course, there were many more variations. To imitate the light of a fireplace, the cinematographer would create a flickering effect. To imitate the light of a handheld lamp, the cinematographer would create a moving light effect. To imitate the light of a handheld candle, the cinematographer would create an effect that moved and flickered at the same time. One unusual variation appears in *Dark Victory* (Edmund Goulding, 1939): as the doctor examines the eyes of Judith (Bette Davis), a carefully focused spotlight moves back and forth, in coordination with the doctor's lighting tool. Cinematographer Ernest Haller dims his key-light to make the light-effect more visible. This may seem like a one-of-a-kind effect, but it is a variation of a technique that can be found in other films, such as King Vidor's *The Citadel* (1938, shot by Harry Stradling Sr.). In other words, it is the creative application of a shared convention.

The ASC was not the only institution to praise the merits of effect-lighting. Studio bosses were also aware of this idea. While Tony Gaudio was working on *The Amazing Dr. Clitterhouse*, Hal Wallis sent a stinging memo to the film's producer, Robert Lord:

> Instead of having sketchy lighting up on the roof and playing it in patches of light and shadow, the whole set has so much general lighting to make it appear that it is being played in broad daylight, and also to bring out the fakey

quality of the set itself. I am really surprised that Gaudio didn't do a better job on this, as the sinister, exciting, mysterious feeling that this sequence should have is missing because of the general lighting of the set. It absolutely takes the kick out of the whole thing. Everybody's walking around and you get the appearance that they are on Seventh and Broadway. There is no feeling of furtiveness, or secrecy, or the necessity of keeping under cover or anything of the kind. Everybody is walking around and the set is brightly lighted, and the whole damned thing to me is lousy.[15]

Wallis is demanding that a nighttime scene be lit with a nighttime effect—especially given the fact that this nighttime scene shows criminals robbing a store. Significantly, the remainder of the scene is filled with striking effects, including some flashlight effects that appear to be illuminated with a powerful bulb placed inside a real flashlight. Apparently, Gaudio did not want to risk inspiring another memo like this one! Wallis probably was not aware of the details of effect-lighting, but he was clearly aware of the functional justifications behind it.

Although most lighting effects are based on conventions, the fact that they are called "effects" suggests that they are supposed to stand apart from the overall lighting of the film. When an image contains a lighting effect, the effect is often the brightest and most densely patterned component of the overall image. This produces a certain unresolved tension surrounding the techniques of effect-lighting. On the one hand, many cinematographers were proud of the unobtrusiveness of their work, with no anxieties about the supposed artlessness of unobtrusiveness. On the other hand, however, they looked at effect-lighting as an opportunity to prove their artistic skills—in the example above, Gaudio seems to be trying to prove a point to his ungracious boss.

Even cinematographers who normally champion the ideal of invisibility could remark on the obvious appeals of effect-lighting. For instance, in the last chapter we saw that Charles Lang was proud to be an unobtrusive artist. Here is the same Charles Lang, on the virtues of effect-lighting:

And above all, if the story will possibly stand it, don't be afraid to go strongly for effect-lightings! They're a sure-fire winner of the average non-photographer's praise (you never saw any cinematographer miss on a mystery or horror picture, did you?) and while you may fail to satisfy yourself on some of the more conventional shots, in details only a cinematographer would notice, if you get in a good sprinkling of really striking effect-lightings and forceful compositions, you'll find the front office, the director, and the stars are all likely to pat you on the back as a rising young artist![16]

The tension is similar to the controversy that surrounded Belasco's lighting effects, detailed in chapter 3. Belasco claimed that his ideal was unobtrusive artistry, but he was accused of sacrificing story in the interest of creating eye-catching spectacle—an accusation that probably had some merit. That tension is obvious in a film like *The Cheat*, which draws self-consciously on the Belasco tradition. It is less obvious in the classical films of the sound period, but we should recognize that it had never completely disappeared.

As in the silent period, the techniques of effect-lighting overlapped with another set of techniques: the genre/scene conventions. Lang mentions that the mystery and horror genres were particularly well suited to the techniques of effect-lighting. This is certainly true, but the conventions were ultimately distinct. A lighting effect could appear in a Bette Davis romance or in a Deanna Durbin musical. To better understand this distinction, we must turn to the genre/scene conventions themselves.

Genre/Scene Lighting

Like the conventions of effect-lighting, the genre conventions developed gradually over time, moving from one-off experiment to recommended standard. By the early sound period, some conventions were so commonplace that ASC publications began to cite them as examples of the everyday artistry of the Hollywood cinematographer. For instance, in his 1930 essay "Painting with Light," Victor Milner mentions three sets of genre conventions:

> In a well-photographed picture the lighting should match the dramatic tone of the story. If the picture is a heavy drama, such as *The Way of All Flesh*, *Lummox*, or *The Case of Sgt. Grischa*, the lighting should be predominantly sombre. If the picture is a melodrama, like *Dr. Fu Manchu*, or *Alibi*, the lighting should remain in a low key, but be full of strong contrasts. If the picture is, on the other hand, a light comedy, like *The Love Parade*, the lighting should be in a high key throughout, for two reasons: first, to match the action, and, secondly, so that no portion of the comedy action will go unperceived.[17]

Milner's genre conventions draw on precedents established in the theater. Indeed, Milner's ideas seem remarkably similar to the idea of Theodore Fuchs, writing about theater lighting in a book from the previous year: "the rule of thumb seems to have been: bright light in full blast for comedy and farce, a dim light for deep tragedy, and all the proportional gradations of lighting for the intervening range of emotions."[18] As in the theater, Milner offers his genre conventions as a distinct category, separate from the effect-lighting conventions. Whereas the primary justification for effect-lighting is a certain kind of realism (we might call it realism-of-detail), the primary justification for the genre conventions is expression: the lighting must match the dramatic mood of the story. This justification is unambiguously aesthetic; Milner writes that mood lighting is "the highest development of artistic photography."[19]

As an indication of the increasing conventionality of genre lighting, we can find similar advice in articles by other cinematographers. Three years later, Charles Lang wrote an important article about diffusion. Lang's article is not exclusively about lighting, but his recommendations are worth mentioning because they overlap so closely with Milner's own. Lang writes:

> An ultra-realistic or melodramatic story, such as *Scarface*, on the one hand, or *Jekyll and Hyde*, on the other, demands a definite harshness and contrast in the photography; clearly, diffusion will be of little use here. The same is true of broad comedy, where high-key lighting must flood every corner of the set, so that no slightest bit of action is lost. The more polished, dramatic comedy and comedy-drama, on the other hand, are usually enhanced by consistent, though slight, diffusion throughout. Most dramas, of course, demand a greater degree of diffusion, while romantic or sentimental plots almost always call for the greatest degree of diffusion of all.[20]

Like Milner, Lang calls for strong contrasts in the melodrama, and bright lighting in the comedy. Lang also adds a new category: the romantic-sentimental film, which calls for soft treatment. However, this does not mean that Lang is proposing a radical new idea. Lang is simply articulating an idea that had already been established as a practical norm.

Before we look at some examples, consider one more passage outlining the basic norms. The point of this redundancy is to emphasize how widespread these techniques had become. Simple guides had become default conventions. The ASC was using repetition to make these rules as widespread as possible. In a 1936 article, longtime ASC president John Arnold writes:

> If we are photographing a heavy dramatic situation, we strive for sombre, "low-key" lightings whose dark tones will heighten the sense of tragedy. If it is a melodrama, strong, virile contrasts between bottomless shadows and intense highlights not only aid in developing a response to rugged action, but etch the action clearly and swiftly to the eye. If the picture is cast in a realistic mood, like *Fury*, harsh, almost newsreelesque photography builds an illusion of reality.
>
> If, on the other hand, the picture is a romance, softer, smoother photography builds subtly to an illusion of idyllic glamour. Lastly, if our picture is a broad comedy, camera and lighting must simply reveal a stage for the comics, without a trace of artifice or artiness, so that not even the smallest gesture, the slightest grimace, will slip by unseen.[21]

Arnold is the first to draw a distinction between the "realistic" film and the melodrama, but the rest of his recommendations follow the predictable pattern: somber lighting for drama, bright lighting for comedy, soft contrasts for romance, and strong contrasts for melodrama. Arnold also puts extra stress on the multifunctional advantages of genre-specific lighting, referring to the normative ideals of illusionism and efficient storytelling, but the primary justification remains an expressive one: the lighting should match the mood of the story.

These lists of conventions would be of little interest if they did not point us to stylistic tendencies in actual films. How did the cinematographers of the ASC put these words into practice?

Melodrama

It is often assumed that low-key lighting was relatively rare in the thirties, appearing mostly in cheap films where the cinematographer did not have to worry about studio supervision. It is true that we can find plenty of melodramatic lighting effects in B-films like *Thank You, Mr. Moto* (Norman Foster, 1938). Some cinematographers even preferred shooting B-films because they provided more opportunity to experiment.[22] However, we should remember that genre conventions could be applied to individual scenes, and not just to entire films.

With this in mind, we can find plenty of examples of melodramatic lighting in A-films, since a crime scene can appear in almost any genre. Cecil B. DeMille's *Cleopatra* (1934) includes a very famous crime scene—the murder of Julius Caesar. Victor Milner lights this murder the way he might light the killing of a two-bit gangster, by lighting the killers from below (fig. 6.9). The Frank Borzage romance *History Is Made at Night* (1937) begins with a robbery scene, which David Abel films with exquisitely dark shadows. In *Peter Ibbetson* Charles Lang films the prison scenes with strong contrasts, and he lights a man from below during a fateful shooting scene. In the murder scene from the drama *Kid Galahad* (1937, directed by Michael Curtiz), Tony Gaudio uses the reliable cast shadow convention. Even a screwball comedy can have a melodramatic scene. In *The Mad Miss Manton* (1938, directed by Leigh Jason), Barbara Stanwyck plays a madcap heiress entangled in a murder plot. The comic scenes are lit for comedy, but the crime scenes show the talent for melodrama that cinematographer Nick Musuraca would later bring to classic noirs like *Out of the Past* (1947). In one scene Musuraca uses a hard, low-placed key-light to create the effect of a woman lit by the flame from a cigarette lighter.

All of these examples are indisputably A-films, featuring multiple stars and major cinematographers. To this list we can add the melodramatic scenes in *The Bat Whispers* (Roland West, 1930), *Mata Hari* (George Fitzmaurice, 1931), *Barbary Coast* (Howard Hawks, 1935), *Anthony Adverse* (Mervyn LeRoy, 1936), *The General Died at Dawn* (Lewis Milestone, 1936), *The Hunchback of Notre Dame* (William Dieterle, 1939), and *The Roaring Twenties* (1939).

Figure 6.9 *Cleopatra* (1934): The murder of Julius Caesar, with the murderers lit from below.

With its use in so many films, we cannot think of melodramatic lighting as an Expressionistic violation of the classical norm. Instead, we should acknowledge that there was no single norm against which exceptions could be measured. The whole point of having a set of genre/scene conventions was to give the cinematographer a variety of choices, each one fully acceptable in the right context. The normative ideal was an ideal of differentiation, resulting in several distinct practical norms, including separate norms for the drama, the romance, the comedy, and perhaps even the "realistic" film.

Drama

The most common term in descriptions of dramatic lighting is "low key." This usage appears frequently in the pages of *American Cinematographer*, and numerous film historians have picked it up. Unfortunately, the term is somewhat ambiguous. Most film scholars use it to refer to high-contrast lighting—that is, lighting with a large difference between key and fill. Milner distinguishes between the somber style of the heavy drama and the high-contrast style of the melodrama. Referring to the latter, he writes, "The lighting should remain in a low key, but be full of strong contrasts."[23] This clearly indicates that "low key" is *not* synonymous with "strong contrasts." Similarly, Arnold associates "dark tones" with the drama, and strong contrasts with the melodrama. A 1934 article from *International Photographer* makes these distinctions even more explicit:

> The "key" of lighting is always an important consideration—high key, in which highlights predominate, for joy, happiness, gaiety, airiness, and delicacy; low key, in which dark shadows and somber grays predominate, for somberness, tragedy, severity, and death; medium key, containing only gray with no extreme highlights and shadows, for fogginess, vagueness, dejection, and impending danger or tragedy; and contrasty key, containing extremes of highlights and shadows, for the weird, mysterious, horrible, and uncertain.[24]

For this writer, low-key lighting has somber tonalities, but it does not necessarily have strong contrasts. Following the associationist logic of mood lighting, "extreme" contrasts are reserved for "weird" moods. The distinction does indeed mark a difference. Contrast refers to the ratio of the brightest tones to the darkest tones. An image that is entirely composed of dark tonalities will be a low-contrast image, in spite of its dominant tonality. To complicate matters further, it is not always easy to describe an image's dominant tonality, because a cinematographer could light the foreground, middleground, and background differently. Here again, the terms "low-key" and "high-key" do not allow us to notice important distinctions. To avoid this confusion, it seems preferable to refer directly to an image's contrast and dominant tonality, noting foreground/middleground/background distinctions when appropriate.

With these caveats in mind, we can turn to the heavy drama. Arnold and Milner both use the term "sombre" to describe this genre's characteristic style. Since they distinguish this from the high-contrast style of melodrama, it seems most likely that they are referring to the dominant tonality of the image, rather than to its level of contrast. This gives the dramatic cinematographer a range of options. The image can be high-contrast or low-contrast, as long as the dominant tonality is dark—not necessarily black, but somber enough to be noticeably darker than the bright images of comedy.

In John Ford's *Arrowsmith* (1931), Ray June uses the dramatic conventions to light the scene when the protagonist's wife dies. In figure 6.10, a kicker outlines Ronald Colman's facial features, but June has flagged off the fill light, casting a shadow over the rest of Colman's face. The lights are perfectly calibrated: the kicker produces a glint of light in Colman's eye, and there is just enough detail to allow us to see his pained expression. The light background serves to make the shadows seem even darker, maximizing their expressive impact.

Figure 6.10 Shadowy lighting for a serious scene in *Arrowsmith* (1931).

It is rare to find a Hollywood film that feels like a "heavy drama" from start to finish. Most Hollywood films include a romance plot, and even a serious film like *Arrowsmith* will include moments of comic relief. Still, we can find somber cinematography in films that are not as consistently glum as *Arrowsmith*. This is because the genre conventions were also conventions for lighting individual scenes. A cinematographer might reserve the somber style for the most serious moments. In Frank Borzage's *Desire* (1936), the overall mood of the film is bright and charming—until Marlene Dietrich's thief realizes that she must tell the man she loves that she has been deceiving him. This is a serious scene, and accordingly, Charles Lang shifts to a somber style. Dietrich stands up into a shadow before exiting the room. The image of her standing in shadow lasts for only a second, but it clearly sets the mood for the scene: a mood of serious drama.

Romance

During the silent period cinematographers would occasionally use the devices of the soft style to set the mood for romance—though this was certainly not the only function of the style. After the transition to sound, most cinematographers abandoned the extremes of the suddenly old-fashioned soft style. Cinematographers began to moderate the technique, applying fewer and fewer layers of diffusion to their lamps and lenses. This process increased as the decade progressed, culminating in the noticeably harder style of the forties. The romance conventions grew less obvious over the course of the thirties, but they remained important. Indeed, the passages from Arnold and Lang suggest that they were becoming more conventionalized, in the sense that all cinematographers were supposed to follow a few basic guidelines.

Consider two examples from Frank Capra's *It Happened One Night* (1934, photographed by Joseph Walker). In figure 6.11 there are none of the dark shadows of melodrama; instead, we have bright highlights mixed with various shades of gray. A frontal key-light is selected to smooth the features of Claudette Colbert's face by minimizing the shadows. A strong backlight provides some highlights, which gain an extra glow from the use of lens diffusion. This glow serves to ease the transition between the brightest part of the image and the darker background. Similarly, the shallow depth of field softens the background itself, allowing the darker grays to gently merge with the lighter grays. The result is an image that embodies the "softer, smoother" ideal mentioned by Arnold.

At first glance, figure 6.12 appears less romantic, since it does not include the blazing backlight. However, notice the way that Walker handles the shadow on Colbert's face. This is not an attached shadow, created by the shape of Colbert's features. This is a cast shadow, created by a flag placed in front of the key-light. An attached shadow would produce modeling, emphasizing the shape of her face. The cast shadow emphasizes the smoothness of her features, while producing a similar nighttime effect. Two other details are worth noticing. First, this is still a low-contrast image, since we can see plenty of detail in the highlights and in the shadows. Second, the edge-line of the shadow is somewhat soft, suggesting that Walker has put diffusion on the lamp providing the key-light. In other words, this image combines lens diffusion and lamp diffusion to produce an extra level of softness.

Figure 6.11 A glowing backlight in *It Happened One Night* (1934).

Figure 6.12 The shadow on Claudette Colbert's face has a soft edge.

We might expect the romance to favor bright tonalities, since the mood is generally positive, as in the comedy. However, while strong blacks are to be avoided, romantic images like this one are still somewhat darker than the prototypical comedic image. There are two possible explanations for this technique. First, romance is commonly associated with night, and a moderately shadowy image evokes the image of night without plunging the characters into a melodramatic abyss. Second, the positive emotions of romance are often mixed with sadness, which calls for an increased level of shadow. Here, the gentle shadows on Colbert's face help to establish the nighttime setting, while evoking a mood of sadness that is appropriate to this image of romantic longing.

The contrast between the romance and the melodrama can be extended to include some observations about the use of effect-lighting. It is too easy to say that some effects were appropriate for the melodrama and inappropriate for the romance. Instead, we should say that these two genres tended to favor different kinds of effects. For instance, the melodrama often employs effects that show a character lit from below. This suits the mood of the genre, because such an unglamorous technique can make the character seem harsh. By contrast, the romance tends to favor effects that add to the pictorial beauty of the film. For instance, Charles Clarke recommended silhouettes and semi-silhouettes for romantic scenes, "as the shadows would suggest a certain amount of privacy and intimacy."[25] He also recommends the use of cast shadows to produce pictorial patterns: "A night exterior, romantic scene is always more convincing if the players stroll through an interplay of light and shade."[26] Similarly, Charles Lang said, "And romantic shadows for romantic photography are extremely necessary. Leaf shadows or whatever. . . . It gives an atmosphere."[27] Arthur C. Miller uses a terrific cast shadow pattern in figure 6.13, a romantic scene from John Ford's *Wee Willie Winkie* (1937). At first glance, this appears to be a rather unclassical moment of pictorial spectacle, but cinematographers had no objection to spectacle when it could help tell the story. Here, Miller uses a beautiful image to enhance the mood of romance.

As in the drama and melodrama, we should not expect to see the romantic conventions remain consistent across the course of an entire film. Instead, cinematographers would often save their most romantic shots for a film's most romantic scene. Just as Charles Lang saved a somber shot for a dramatic moment in *Desire*, he saves his softest effects for the most romantic scenes in the same film. Here again, the genre conventions function as conventions of differentiation. The cinematographer

Figure 6.13 A beautiful cast shadow pattern for a romantic moment in *Wee Willie Winkie* (1937).

proves his skill by varying his technique in subtle ways, to mark the transition from drama to romance.

This rule of differentiation applies to all the genres, even the genre that was the least visually interesting: the comedy.

Comedy

After the harsh shadows of the melodrama, the somber tones of the drama, and the pictorial beauty of the romance, it is a bit of a letdown to consider the flat images of comedy. Figure 6.14 shows a typical shot from *Horse Feathers* (Norman McLeod, 1932). Ray June was an accomplished cinematographer who would finish his career with three Academy Award nominations to his credit, but his photography for this Marx Brothers vehicle is, to borrow a phrase from John Arnold, "without a trace of artifice or artiness." It is tempting to add "without a trace of effort," since it appears that June simply turned on all his lights and hit the switches on the cameras, allowing Groucho, Chico, and Harpo to run all over the stage. While it was standard practice in most genres to put additional light on the foreground, no such subtleties are allowed to grace this production.

Still, we must not underestimate the challenges involved in achieving such an apparently artless look. In addition to the herculean effort involved in resisting the temptation to cast a little shadow here and add a little sparkle there, June had to face the practical problem of keeping the light even across a large space. The narrative requires an overall high-key style, since almost every corner of the room will have been used by the

Figure 6.14 Flat illumination for the Marx Brothers in *Horse Feathers* (1932).

end of the scene: to the left, a succession of visitors enters through a door; in the background, Groucho places a lamp near the window; and to the right, Harpo heaves a block of ice outside a window. Making matters even more difficult, the Marx Brothers films always required multiple cameras, since they could never predict what the brothers were going to do.[28]

The practical demands of the comedy style can also be demonstrated by an example from the W. C. Fields comedy *It's a Gift* (Norman McLeod, 1934). Henry Sharp had the task of shooting this film—the same Henry Sharp who had photographed such elegant images for Maurice Tourneur in *Lorna Doone* (1922). An extended comic sequence showcasing Fields's failure to fall asleep starts at night and ends in the morning. The morning shot is predictably bright; what is surprising is that the night shot is very similar. Instead of adding more shadows and contrasts (a conventional way to signify night), Sharp dims almost all of his lights equally, producing a nighttime image that is still illuminated over a large area. The nature of the scene requires this "flooded" style, since the gags are occurring all over the set.

As in the silent period, the discourse of Art put comedy cinematographers at a disadvantage. If the job of the comedy cinematographer is, in Arnold's words, to "simply reveal a stage for the comics," then his task is ultimately a matter of mechanical reproduction. He has failed to participate in the "mechanics-to-artists" trajectory that was one of the defining narratives of the ASC. Many cinematographers admitted it was their least favorite genre. Milner described the lighting style for broad comedy as "more conventional."[29] James Wong Howe wrote that he preferred low-key lighting.[30] D.P.s may have referred to cinematography as "painting with light," but the artistry was more accurately found in the shadows.

Cinematographers could take heart in the fact that the "flooded" style was reserved for broad comedies, such as the films of the Marx Brothers or W. C. Fields. Romantic comedies (including musical comedies) could take advantage of the wider range of options available in the romance. In the Astaire-Rogers films, cinematographer David Abel may be willing to flood the stage for a musical number, but he wisely chooses not to shoot Ginger Rogers like Groucho Marx. Instead, he employs the more gentle gradations typical of the female glamour shot.

More important, some cinematographers attempted to integrate the comedy conventions into the larger discourse of expression. Since comedy is generally considered a "joyful" genre, a cinematographer could describe the high-key style as "bright" or "brilliant," thereby linking the style to the mood. Furthermore, the choice to flood a scene with light was still a choice, tailored to the specific needs of the narrative. Like his peers working in the genres of melodrama, drama, and romance, the comedy cinematographer could claim to be an artist with a special skill in interpreting the story.

Compositional Lighting

In a 1930 article on "Composition in Motion Pictures," cinematographer Daniel Clark links composition to the mechanics-to-artists narrative:

> It is possible to arrange a scene so that the vision is led or guided to any desired spot, and held there, suspended and waiting, even though action is taking place at another place in the same field of vision. It is the skill exhibited in doing this that differentiates the artist from the "crank turner." By composing the set and actors on it, the skillful cinematographer can make identical action difficult or easy to follow. The cinematographer's ability to compose is often the means of clarifying otherwise difficult parts of a story.[31]

A crank turner merely captures whatever is in front of the camera. An artist structures the image, making some parts more salient than others. Cinematography is seen as a creative act because the cinematographer makes distinct choices to enhance the clarity and overall readability of the image. While the director and art director rule the actors and the look of a set, the cinematographer's dominion is to control the final look of the image projected on the screen.

During the silent period cinematographers had developed a variety of strategies for directing the viewer's attention, such as selective focus and the use of theatrical scrims to diffuse the background. Some of these techniques quickly fell out of favor with the advent of talkies. Because the transition to sound gave privileged status to the discourse of fidelity, heavy diffusion and theatrical scrims came to be seen as overly obvious aesthetic tricks that detracted from a film's realism. Still, several silent strategies continued to be important. For instance, a 1934 article in *International Photographer* affirms the continuing importance of selective focus: "In professional pictures the background of some action is usually kept unobtrusive by throwing it out of focus."[32] As is well known, cinematographers like Gregg Toland and Arthur Miller were beginning to experiment with deep-focus photography during the latter part of this period. Still, it seems fair to say that selective focus remained the default option. In a typical close-up, a cinematographer would open the aperture and reduce the depth of field, keeping the attention on the star.

In the hands of a skillful D.P., lighting could be used to establish extraordinarily subtle gradations of emphasis. Writing in 1931, James Wong Howe explains his method of lighting a scene, pointing out that guiding attention is his first priority:

> Every scene has its centre of interest. It may be the face of one actor, or of several; it may be merely a part of a face; it may be a hand or foot; it may even be some inanimate object—a letter, a pistol, a key, or a dropped handkerchief. Whatever it is, it is that feature which for the moment is most important in advancing the story. . . . If you build your composition on that, and then build your lighting on the composition, why—there's your picture![33]

Lighting gives hierarchy to the space, giving graded emphasis to the various components of the scene. The structure of the hierarchy is governed by the logic of the story. This gave the cinematographer a range of options. Depending on the needs of the scene, a cinematographer could construct a complex hierarchy, a simple hierarchy, or anything in between.

In figure 6.15, from *Peter Ibbetson*, cinematographer Charles Lang constructs a four-level hierarchy. The strongest light is on the boy, who rivets our attention. (Director Henry Hathaway's blocking also helps, as the boy is in the foreground, near the center.) We might expect the next strongest light to be on the woman walking through the mid-ground, but Lang has shrewdly realized that this woman is less important than the empty chair in the background—a chair that represents the boy's dying mother. Lang puts a secondary key-light on the chair, and gives a weak side-light to the woman—strong enough to create the sense of roundness, but not so strong as to distract our attention from the boy and chair. The lowest level of the hierarchy is given to the room itself, with just enough general illumination to maintain a minimal sense of time and place.

Perhaps the most common strategy was the two-part hierarchy, putting more light on the foreground than on the background. John Arnold, the head of the cinematography department at MGM, favored this as a default strategy: "When tonal values of characters and background are much the same, light must be laid heavier on the characters. From a third to a half more light should be in the foreground than in the background."[34] While this advice initially appeared in *AC*'s amateur section, Arnold's own subordinates at MGM followed the suggestion routinely. Figure 6.16, from *Anna Karenina*, is a particularly subtle example. Here, the key-light emphasizes Anna (sitting) and Vronsky as they look at each other with interest. The distance between Anna and Vronsky is significant, and Daniels has to make sure that the three figures in the background will not become a distraction. He lights them perfectly—a little more light, and the figures would draw too much attention; a little less light, and the unmotivated darkness itself would become noticeable.

If the foreground/background strategy was simpler to execute than the multiple degrees of emphasis strategy, the simplest option of all was to light the entire space evenly. This failed to produce a hierarchy, but it was still a creative choice, since certain scenes called for such a uniform treatment. Musical numbers provide the most obvious example: since the star may dance all over the stage, the best way to keep the attention on her is often to light the entire set. The same idea justified the flat, high-key lighting of the Marx Brothers comedy.

Figure 6.15 *Peter Ibbetson*: Lighting directs our attention to the foreground and background, while keeping the middle-ground unobtrusive.

Figure 6.16 *Anna Karenina*: The foreground is brighter than the background.

Even before directing the spectator's attention to a salient point, a cinematographer had to accomplish an even more basic step: give the image an adequate exposure. This may seem so obvious that it is not worth mentioning, but Lucien Ballard once tried to test the limits of this rule by using five seconds of total blackness for dramatic effect. Years later, he talked about the uproar it produced:

> That night, Harry Cohn saw the rushes, and he called me on the phone to fire me. Fortunately, I was out that night and couldn't be reached, but I saw him the next day and he told me that if he'd gotten through to me I would have been fired. "I pay all those actors," he said. "What's the idea of having them in total blackness for ten minutes?" I said, "It was only for the count of five." He said, "I want to see my actors at all times!"[35]

The convention requiring adequate exposure in every shot demarcated the perimeter beyond which not even the most daring cinematographer was allowed to tread.

A skillful cinematographer knew how to combine the strategies of composition with other techniques, but composition often took the first priority. Even James Wong Howe, a strong advocate for the value of motivated light sources, recommends motivating the lights only after ensuring that they work within the context of composition. Continuing the passage quoted above, he writes:

> In building up your lighting this way, you first light this salient point (or points). Get enough front and side light to give good definition, and enough back and top light to give the necessary modeling. Then add enough light elsewhere in the picture to bring out such or the other details as you want, the

way you want them brought out. Then check the whole carefully to see that the result is natural. In almost every instance there should be one definite source from which the light should appear to come.[36]

Guiding attention seems to be Howe's initial priority: his first move is to put a key-light on the most salient plot point. Next, he secures the illusion of roundness by working with fill lights and backlights. Then he returns to the problem of guiding attention, making sure that secondary story points also receive a little extra light. Only then does Howe consider the problem of motivation, tweaking the light to suggest that it is produced by a light source within the fictional world.[37] Here, Howe's desire for realism converges neatly with the goal of guiding attention. If a certain light were to lack motivation, spectators might be distracted by it. This would be counterproductive, since the reason the light was there in the first place was to direct the spectator's attention to something else.

Some cinematographers chose a slightly different approach. Whereas Howe sets the highlights first and the general illumination second, Victor Milner recommends setting up the general illumination first and then adding the highlights.[38] Either way, the cinematographer needed to balance the two to achieve the illusion of roundness. In his major essay "Painting with Light," Victor Milner places particular emphasis on this function. Comparing the cinematographer to a painter, he writes:

> The problem of both is the production of an illusion of roundness and depth on a flat surface. While the physical means employed are different in the two forms, in the final analysis we can see that the essential tool is the same; for, both on the canvas of the painter and the screen of the cinematographer, these effects, together with those of mood and character, are secured by the careful manipulation of light and shade.[39]

In a time of technological change, Milner relies on familiar rhetoric to maintain the stability of the cinematographer's public identity. The screen is a canvas; the cinematographer is an artist.

Indeed, Milner does more than simply reaffirm the importance of the illusion of roundness. He helps to reestablish the norms by offering detailed practical advice on how to construct this illusion. First, Milner suggests that the cinematographer can use contrasting lighting for each major plane of the picture. For instance, if a shot contains a foreground, a middleground, and a background, then the cinematographer has two options: 1) darken the foreground, add light to the middleground, and darken the background; or 2) add light to the foreground, darken the middleground, and add light to the background. Either way, the cinematographer should attempt to alternate his lighting scheme to distinguish among the three planes. This strategy adapts and extends the *repoussoir* technique developed in the silent period. Figure 6.17 shows an image from *Trouble in Paradise*, a 1932 Lubitsch film shot by Milner. The bed railing in the foreground is mostly in shadow; Herbert Marshall occupies the brightly lit middleground; and the large case in the right background is obscured in shadow. By using contrasting lighting, Milner emphasizes the presence of three distinct planes. Second, Milner proffers that the cinematographer can create depth through the careful placement of highlights. This technique is particularly useful on curved surfaces. In the image, notice how the highlights on the finial in the foreground emphasize the roundness of its shape. Third, Milner recommends highlighting the walls of the set. Assuming the walls are not completely flat, judicious lighting can emphasize their texture. Finally, Milner suggests the cast shadow as a technique for clarifying spatial relationships, since it can give information about shape and distance.[40]

Milner does not mention backlighting as a technique for enhancing the illusion of depth, but it draws on two of his general principles. By adding light to an actor's hair, a cinematographer can separate the actor from the background. Since the actor's head is also a type of "curved surface," the backlight helps express shape as well.[41] All of these techniques would be useful composition conventions throughout the thirties.

Another important function of composition was the creation of pictorial beauty. Given that cinematographers were now fully committed to the ideal of invisibility, we might expect cinematographers to have abandoned this particular ideal. However, even if we bracket off pictorialists

Figure 6.17 Multiple planes of depth in *Trouble in Paradise* (1932).

like Josef von Sternberg, this restriction on pictorial beauty was not absolute, for several reasons. First, we should remember that a picture does not have to be eye-catching to be beautiful. Most Hollywood images will meet minimum standards of order and balance. Second, pictorial beauty was perfectly appropriate for some genres, such as the romantic drama, especially if the drama had an exotic setting. Third, certain shots benefit from pictorialism more than others. For instance, Charles Clarke suggests that an establishing shot provides a good opportunity for pictorialism, since an establishing shot usually appears at the beginning of a scene, before the drama has begun.[42] This suggests an alternation model, with the film constantly switching back and forth between a pictorial mode (in the establishing shots) and an invisible mode (in the coverage).

Fourth, some of the most common pictorial strategies were designed to add a dash of beauty without distracting from the story—or, at least, without distracting from the story too much. For instance, many cinematographers used the window-pattern technique to add a dash of pictorial beauty to an otherwise mundane composition. Cinematographer George Folsey enjoyed working with light walls because he thought they were "a perfect canvas on which to paint light effects." He advises:

> The first step in lighting such a set is to take advantage of everything which can logically be used to cast either shadow or light patterns on these walls. Very frequently, you will find a prop—a flower, a chair, a statuette, or the like— which can be utilized for this shadow-casting. Similarly, you will usually find lamps, windows, doors, etc., which can be equally useful for casting or legitimizing high-lights. If you can balance your lighting properly, you can get beautiful, luminous shadows and fine, soft high-lights, with a delightful range of half-tones between.[43]

For Folsey, the primary justification for these effects is not realism—it is pictorial beauty. Once we begin to look for them, we can find beautiful patterns throughout the classical cinema. George Barnes won the Best Cinematography Oscar for *Rebecca*, which is filled with stunning window patterns. For a less celebrated example, consider figure 6.18, from *Three Smart Girls*. The powerful light blasting through the skylight adds a dynamic diagonal to the composition, which looks almost like a Charles Sheeler painting. Looking at a shot like this, it is no surprise to learn that an *American Cinematographer* article about Valentine was titled "Valentine's Technique is Vivid and Modern."[44] Light effects still had the power

Figure 6.18 A dynamic window pattern in *Three Smart Girls* (1936).

to produce an invigorating pictorial spectacle, no matter how conventional they had become.

Indeed, casting a shadow on the back wall became a routine composition convention, independent of effect-lighting considerations. Many cinematographers throw a shadow on the back wall whether the shadow is motivated or not. In figure 6.19, from *Bachelor Mother*, a strong side-light duplicates the pattern of the banister. Cinematographer Robert de Grasse could have motivated this as an effect, but he does not even bother. Instead, we can say that cinematographers had developed a new rule: If you see a banister, shine a light through it. This rule captures the ASC's complicated attitude to pictorial beauty. On the one hand, cinematographers routinely created beautiful patterns. On the other hand, they kept them, quite literally, in the background.

As with the other sets of conventions, cinematographers responded to the challenges of the transitional period by reasserting the norms they had worked so hard to establish during the silent period. The discourse

Figure 6.19 *Bachelor Mother* (1938): Even a screwball comedy can use a banister pattern.

and practice of composition still relied on the functions established during the silent period: narrative clarity, illusionism, and a judicious use of pictorial beauty.

◎ ◎ ◎

This chapter has emphasized the coherence of classical Hollywood lighting. Cinematographers had developed a complex set of conventions, and they worked hard to make those conventions work together. Each of these conventions had evolved independently, providing practical solutions for particular problems. However, perfect convergence was not always possible. The style of the classical Hollywood cinema had become so complex that it forced a new problem to the foreground: following one rule might require the cinematographer to break another. In the next chapter, I will take a closer look at this problem, and examine some of the ways that cinematographers tried to solve it.

7 The Art of Balance

In the climactic scene of *After the Thin Man* (W. S. Van Dyke, 1936), the elegant detective Nick Charles (William Powell) reveals that the apparently friendly fellow, David, played by Jimmy Stewart is actually a murderous psychopath. Right before the revelation of his crimes, David receives gentle glamour treatment from a diffused key-light in a front-side position (fig. 7.1). The film then cuts to Nick as he delivers his most damning piece of evidence. When the film cuts back to David, his soft key-light has become a hard top-light (fig. 7.2). At the moment of the audience's realization, the lighting shifts to emphasize the contortions on the killer's face as his insanity comes to the surface.

In one way, this example seems to be thoroughly conventional. The right-mood-for-the-story theory encourages cinematographers to vary the lighting to suit the changing moods of the story. The conventions of figure-lighting encourage cinematographers to use lighting as a tool for characterization. Here cinematographer Oliver Marsh manages to follow both of these conventions at the same time. However, it is clear that Marsh has broken as many rules as he has followed. He has failed to motivate the lighting change as a plausible effect. As a result, the lighting change seems to be a clear violation of Hollywood's principle of continuity. We might say that Marsh has come up against the problem of conflicting conventions.

So far I have argued that the discourse of lighting grew more classical after the transition to sound, as cinematographers embraced ideals like the illusion of presence and the commitment to storytelling. The increasing homogenization of the discourse led in turn to the increasing conventionalization of the style. There was a certain irony in this development:

Figure 7.1 *After the Thin Man* (1936): Jimmy Stewart, before we learn that his character is the murderer.

Figure 7.2 A moment later, after we have learned the truth. Now he is lit by an unflattering top-light.

there may have been more rules, but no cinematographer could follow all of them at the same time. Instead, a certain amount of idiosyncratic interpretation was required. Different cinematographers solved the problem of conflicting conventions in different ways, some more classical than others. Some cinematographers would take a classical approach, balancing the functions of storytelling, realism, glamour, and pictorial quality. Other cinematographers would push the limits of the style, emphasizing one or two artistic functions at the expense of all the others.

Although they did not have a specific term for it, the cinematographers of the thirties were well aware of the problem of conflicting conventions. In a 1936 article, longtime ASC president John Arnold explains:

No picture is going to sustain exactly the same mood throughout all of its many hundred scenes and set-ups. Even in the heaviest drama we have moments of romance, or of robust humor. If these scenes are photographed in exactly the same style as the rest of the picture, they would lose much of their dramatic value; on the other hand, if they are photographed for themselves alone, without consideration for the basic mood of the whole story, they would make visual changes which would distract audience-attention and make the picture seem spotty and disjointed. Properly balancing between the individual requirement of the scene itself and the sustained mood of the production as a whole gives the Cinematographer a greater problem than any faced by Director or Actor. . . . Now add to this problem the everchanging one of keeping the players consistently looking their best, and you have a problem only Einstein's mathematics can fully express.[1]

A heavy drama may call for dark tonalities, but the rules of composition might require a brighter style, to ensure the easy legibility so prized by the Hollywood narrative tradition. A melodrama might necessitate minimal diffusion, but the rules of figure-lighting could mandate a softer style, to suit the needs of a female star. A romance might benefit from an additional backlight, but the rules of effect-lighting might require light from below, to imitate the look of a particular setting. Since Arnold is indulging in some typical P.R., we do not need to accept his claim that the cinematographer's task was the hardest of all. But we can certainly accept his claim that the task was much harder than it looked.

The simplest way to solve the problem of conflicting conventions was to sacrifice one convention in favor of another. Figure 7.3 shows an example from *The Adventures of Sherlock Holmes* (Alfred Werker, 1939). Notice the implausible backlight creating a halo around the actor's head, even though the supposed light source is clearly in front of him. Here, cinematographer Leon Shamroy is facing a conflict between the effect-lighting conventions (which would mandate the rejection of a backlight)

Figure 7.3 A complicated lighting scheme from *The Adventures of Sherlock Holmes* (1939).

and the figure-lighting conventions (which would mandate the inclusion of a backlight). Shamroy opts to sacrifice the consistency of the former in order to secure the advantage provided by the latter—namely, the spatial advantage of separation, as the backlight allows the actor to stand out distinctly from the background.

James Wong Howe describes a different sort of problem that requires a different kind of sacrifice: "I've had problems with fakery in films. In *The Thin Man*, the director wanted shadows because it was a mystery, too many shadows. Often, I didn't see where a light was in a room to make a shadow, but he'd say, 'There's a blank space on the wall, let's fill it,' and I'd make him a shadow. I threw away naturalism on that one, unfortunately."[2] At the director's request, Howe sacrificed the integrity of the effect-lighting to follow a popular genre/scene convention, using cast shadows to set the mood for mystery.

As Howe's words suggest, the sacrifice was not the ideal solution to the problem of conflicting conventions. Sacrificing one rule to obey another was a flawed solution, because it failed to accomplish the multifunctional ideal. A more effective solution to the problem of conflicting conventions was the compromise. In an article on the use of diffusion, cinematographer Charles Lang discusses the following problem: How much diffusion should you use when photographing a scene featuring a man and a woman? The figure-lighting conventions would dictate using less diffusion on the man, and more diffusion on the woman. However, this would produce abrupt changes in style in the middle of the scene, thereby violating another rule: the rule that style should not call attention to itself. Referring to an unnamed film, Lang proposes the following solution:

> To treat such a sequence properly, it would be the cinematographer's problem, first of all, to determine whether the sequence, dramatically speaking, was the man's or the woman's. In this instance, it was the latter: accordingly, he would have to modify his technique and use some diffusion on the man, so that the sequence might remain visually uniform; if necessary, he could light the man a bit harder than usual, to compensate. If, on the other hand, it were the man's sequence—that is, if his action, reaction, and dialog took precedence—the best treatment would be to reduce the diffusion upon the woman—possibly compensating by softer lighting—and bring the sequence into visual coordination this way.[3]

To suit the demands of continuity, Lang could have advised using the same amount of diffusion for both men and women. Instead, he suggests a more complicated solution, involving an increase in lens diffusion and a decrease in lamp diffusion. This is a compromise solution. On the negative side, this solution would not produce perfect continuity, since there would still be small differences between the man and the woman. Nor would it produce perfect characterization, since the man would not be as hard as possible, and the woman would not be as soft as the cinematographer might like. However, this solution is superior to the sacrifice, since it accomplishes the ideal of multifunctionalism, preserving a reasonable amount of continuity while maintaining some subtle distinctions between the man and the woman.

A scene from *Dark Victory* (Edmund Goulding, 1939) provides another example of the compromise. For most of the film, cinematographer Ernest Haller lights Bette Davis with a strong frontal key. This is one of the routine ways of lighting women, since it provides glamorization by smoothing out the facial features. The mood shifts decisively toward tragedy in the second half of the film. In one of the most serious scenes, Haller switches to a different technique, keying Davis from the side (fig. 7.4). In doing so, Haller follows two conventions: the genre/scene convention calling for somber shadows in serious scenes, and the effect-lighting convention calling for motivated shadows in scenes with table lamps. (In a previous shot, Haller has established the presence of a table lamp, now offscreen right.) However, we cannot say that Haller has simply sacrificed the norms of figure-lighting to produce a moody effect. The shadow on Davis's face is not as dark as the genre/scene or effect-lighting conventions would require. Haller has added enough fill light to keep the gradations relatively smooth. He has also employed various other techniques associated with female stars: the subtle use of lens diffusion, the

Figure 7.4 Bette Davis lit from the side in *Dark Victory* (1939).

soft-focus background, and the unmotivated backlight on her hair. The result is an efficient compromise, maintaining Davis's glamour while establishing an expressive and realistic atmosphere.

In figure 7.5, from *Sullivan's Travels* (Preston Sturges, 1941), John F. Seitz strikes a similar compromise by using a slightly different technique. In this shot, the Girl, played by Veronica Lake, mourns the (falsely reported) death of Sullivan. Whereas Haller had produced shadows on Davis's face by moving the key to the side, Seitz produces shadows on Lake's face by putting a net over the key, which is in a front-right position. In so doing, he appears to violate a basic composition convention: "Put the strongest light on the most important detail." Here, the most important detail—Lake's face—is in shadow. However, this solution is actually a remarkably elegant compromise, accomplishing a variety of functions at the same time. First, Seitz uses a somber style to convey a mood of sadness, following the principles of genre. Second, Seitz has created an effect, suggesting the appearance of sunlight peeking through a window with the shade drawn down. Both of these functions could have been intensified with darker shadows, but Seitz has kept the shadows fairly bright by using a high level of general illumination. This allows him to minimize the contrasts on Lake's face, thereby adhering to the principles of figure-lighting. Finally, Seitz has preserved the spirit of the composition conventions, even if he has departed from the normal rule. There are two bright highlights in the shot: the backlight just above Lake's face, and the light on her hands, just below her face. This means that the spectator's attention will still be drawn to Lake's face, even if it is not the brightest component of the shot. In all, this shot accomplishes a remarkable variety of functions— guiding our attention, setting the mood, suggesting the time of day, and enhancing the glamour, all in unison. In order to appreciate the elegance of this solution, we have to understand that Seitz cannot have everything. To increase the compositional clarity, he might add more light to Lake's face, but this would diminish the effect of a pulled-down shade. To intensify the mood, he might add more shadows, but this might reduce the glamour. As it is, the solution is a compromise—a very successful compromise that balances a range of competing demands.

These strategies both involve decisions about the lighting of individual shots. Lighting an entire film could pose similar problems. If a romantic film has two scenes of melodrama, how much should the style change during the melodramatic scenes? If a glamorous star appears in three scenes with lighting effects, how much should the lighting depart from the glamour norms to accommodate the effects? When considering the

Figure 7.5 A shadow is cast over Veronica Lake's face as her character grieves in *Sullivan's Travels* (1941).

lighting strategy for an entire film, it is not enough to note that a cinematographer is choosing a certain sacrifice or crafting a particular compromise. We must also notice *when* the cinematographer is using such a technique.

For instance, John Ford's *The Whole Town's Talking* (1935) is a broad comedy with melodramatic subject matter. When the meek office clerk Jonesy (played by Edward G. Robinson) meets his gangster look-alike Mannion (also played by Edward G. Robinson), cinematographer Joseph August faces a difficult choice. Since this is a major turning point in the melodramatic plot, he could choose to set an appropriately menacing mood by covering the characters in dark shadows. This would set the right mood, but it would also produce a significant disadvantage: too many shadows would cloak the nuances of Robinson's hilarious dual performance. Alternatively, August could choose to use the bright illumination that dominates most of the film. This would also produce a disadvantage, since it might rob the melodramatic scene of its dramatic punch. August's solution is to start the scene with a strong dose of melodrama and then shift to a more moderate style for the remainder of the scene. When Jones first sees Mannion, the gangster is lit with a hard light coming from a right-low position. August cuts the key-light with a flag, producing a mysterious shadow on Mannion's face. For this shot, the mood is decisively melodramatic. A few moments later, Mannion walks into the foreground. The style shifts considerably, as a lamp in the foreground motivates the use of more traditional lighting on Mannion's face. The new lighting scheme allows us to appreciate Robinson's performance. If the first shot is a sacrifice (all melodrama, no comedy), then the second shot is a compromise (some melodrama, some comedy). Here the important point concerns the timing of these decisions. When the dramatic emphasis is on the shocking appearance of the menacing Mannion, August employs a heightened version of the melodramatic style. When the dramatic emphasis shifts to the amusing details of Robinson's performance, August moderates that style, introducing the brighter elements of the comedic style. We can call this the strategy of selective emphasis. Over the course of a scene—or over the course of an entire film—the cinematographer must decide when to use a strong version of a particular style, and when to use a more moderate approach.

Joseph Walker, who shot most of the A-list films at Columbia, was a master of selective emphasis. *Mr. Deeds Goes to Town* (1936) and *Mr. Smith Goes to Washington* (1939) are both comedy-dramas directed by Frank Capra. A casual look at the films might lead an observer to conclude that Walker has simply opted for a flat, unimaginative, neutral style, favoring the comedy at the expense of the drama. However, a closer look would reveal that Walker has simply saved his dramatic shadows for a few carefully selected moments—moments of quiet sadness. A key scene in *Mr. Deeds* shows Deeds learning that his damsel in distress is actually a reporter. Walker photographs this moment in a high-key style, allowing us to see the emotions register on Gary Cooper's face (fig. 7.6). Deeds then walks behind a column. For this quiet moment, Walker uses the column to motivate a gentle shadow on Deeds, expressing his solitary sadness (fig. 7.7). In typical Capra fashion, things soon go from bad to worse, with Deeds winding up incarcerated in an asylum. This is the moment Walker chooses for his most dramatic effect, nearly silhouetting Gary Cooper against a window. This shot is even more remarkable given the location: it is not easy to get such dark tonalities into a shot of a sunlit hospital

Figure 7.6 *Mr. Deeds Goes to Town* (1936): A serious scene begins in bright light.

Figure 7.7 A moment later, Gary Cooper steps into a shadow.

room. The result is an intensely expressive image, fully communicating the gravity of the situation.

Walker uses a similar strategy in *Mr. Smith Goes to Washington*. He photographs the bulk of the film in a high key, saving his deepest shadows for a few significant emotional low points. The film features a romance plot and a political plot. In the romance plot, the major low point comes when Saunders (Jean Arthur) quits her job in Smith's Senate office (Jimmy Stewart), with the intention of marrying Diz (Thomas Mitchell). In the political plot, the major low point is when a crestfallen Smith packs his bags and visits the Lincoln Memorial one last time, an image that previously symbolized Smith's naive idealism. For both of these scenes, Walker shifts his style to a slightly darker tonality. This is a very clear example of the strategy of selective emphasis. Most of the film is bright, but Walker casts a romantic low point and a political low point in somber shadows.

Walker also knew how to save his prettiest shots for romantic high points. In Capra's *Lady for a Day* (1933), he introduces a romantic scene with a stunning image that would not look out of place in a Wong Kar-Wai film: drops of water, falling in the foreground, soften and make abstract an image of two lovers. Walker also uses selective emphasis in Richard Boleslawski's romantic comedy *Theodora Goes Wild* (1936). In this film there are two scenes involving kisses. Walker photographs the first one with the bright tonalities that dominate the bulk of the film. A viewer who stopped watching the film at this point might conclude that Walker was simply ignoring the genre/scene conventions for this assignment. A viewer who continued watching the film, however, would realize that Walker was saving his skills for the second kissing scene. This second scene is unquestionably more important to the narrative—it is the culmination of Theodora's attempt to ruin the marriage of Michael Grant. For this crucial moment, Walker uses a mixture of grays, with few whites and few blacks. The result is an image with the gentlest possible gradations.

The studio context may help explain why Walker favored the strategy of selective emphasis. Walker worked at Columbia, which could not afford spectacular imagery for every single shot. Around this time Lucien Ballard was shooting quickies at Columbia. According to Ballard, "If you had perfection in every shot, you'd lose the story, and besides, the photography wouldn't mean anything—but if you had one or two great shots, say one at the beginning, a highpoint in the middle, and one at the end, those shots would stay in people's minds."[4] Selective emphasis is a particularly elegant solution to the problem of conflicting conventions, because the solution brings an added advantage. As artists have known for some time, using a device selectively can actually enhance its expressive force.

Still, the cinematographer must apply this technique with care. Most Hollywood films are filled with dialogue scenes, and these work better when we can see the actors. It is significant that some of the examples mentioned above involve relatively quiet moments. The skillful cinematographer will save his shadows for the most somber moments, but he will also save them for the moments when they will be least damaging to the film's other needs. If a cinematographer saves his darkest shot for the most dramatic dialogue scene, he runs the risk of making an important moment hard to see, wreaking havoc on the goal of lucid storytelling.

For instance, consider this scene from William Wyler's drama *Dodsworth* (1936, shot by Rudolf Maté). Sam Dodsworth (Walter Huston) has just had a fight with his wife, Fran (Ruth Chatterton), about their vacation; she wants to stay in Europe, while he is ready to go back home to America. It is a serious dramatic scene, but most of the discussion takes place in the brightly lit, well-decorated space that was typical of a Sam Goldwyn film. Toward the end of the fight, Sam sits down quietly, while Fran continues to talk. Fran is still brightly lit, but Sam is now sitting in shadow (fig. 7.8). A few moments later, Sam stands up and exits into the next room, which is much darker than the bedroom. Sam walks through the shadows to the other side of the room. There he turns on a lamp and begins to look through the newspaper. After a beat, he picks up the phone and orders a single ticket back to the States, revealing that he has agreed to Fran's request to give her some time in Europe without him.

In this scene Maté skillfully balances two competing priorities. The genre/scene conventions encouraged cinematographers to use darker

Figure 7.8 *Dodsworth* (1936):
Only one character is placed in
shadow.

tonalities for serious scenes, but it could be argued that this is exactly the
wrong time to use those tonalities here. If the scene is so important,
shouldn't the cinematographer make every detail visible? It is one thing
to encourage the audience to feel the scene, but wouldn't most people
rather *see* it? This question is particularly pertinent in a film, like *Dodsworth*, that foregrounds the acting ability of its stars. Maté's solution is
simple: he saves the shadows for moments that are both somber and
quiet. When Huston sits down in silence, he is shrouded in shadow,
while Chatterton, finishing her speech, remains in the light. Huston's
silent walk to the telephone is also played in darkness, but Maté shifts up
to a compromise position for the scene's final moments, when Huston
turns on a lamp and shifts the dominant tonality to a lighter shade of grey.
With great skill, Maté applies the principle of selective emphasis to the
moment-by-moment development of a scene.

The strategy of selective emphasis had another advantage. Above and
beyond its benefits to storytelling, it also contributed to the overall pictorial quality of a film, making it more visually pleasing. In his 1964 textbook, former Fox cinematographer Charles Clarke notes, "Naturally a
film from beginning to end in one photographic style would be dull and
uninteresting. To make the production stimulating and vibrant, the photographic approach should be changed from sequence to sequence,
within the realms of believability and realism."[5] Clarke goes on to say that
storytelling concerns should dictate lighting changes. However, that
should not prevent us from noticing that pictorial pleasure could be a
benefit by itself, independent of narrative concerns.

While selective emphasis often provides a brilliant solution to a difficult problem, the technique carried its share of risks. A drastic shift in
style might call attention to itself, thereby hurting the story rather than

helping it. In the early thirties, William Stull wrote a regular column for *American Cinematographer*, offering commentary on the cinematography of recent films. As we might expect from such a biased source, the comments are overwhelmingly positive. One of the few "mistakes" that Stull was willing to criticize was the abrupt shift in style. Reviewing the 1932 film *Society Girl* (shot by George Barnes), Stull writes, "He has achieved many individually fine scenes, but has erred in attempting to set too widely divergent photographic moods for the two principal characters—a society girl and a prize fighter. The moods that he has set for these two opposing ideas are in themselves excellent, but when used together, and intercut, they make the picture as a whole unpleasantly uneven and spotty."[6] There are two potential problems here. First, coherence is often thought to be a positive feature in an artwork, and Stull seems to value it for its own sake. Second, a conspicuous lack of unity might draw attention away from the story, diminishing its emotional impact. The mood conventions should intensify the story's emotional impact, not compete with it.

In the same issue, Stull had much kinder words for William Daniels's work on Edmund Goulding's *Grand Hotel* (1932):

> But it is in the less obvious phases of his work that Mr. Daniels has reached his greatest heights. "Grand Hotel" strikes a wide variety of emotional and dramatic moods as it unfolds its story, and the plot-construction requires a great deal of intercutting of these. The manner in which Mr. Daniels has met this artistic problem has never been surpassed, and seldom equaled. It is difficult enough to sustain the simpler dramatic moods of an ordinary feature—and so, Mr. Daniels' achievement in sustaining the photographic uniformity of his picture while matching the infinitely varying emotion keys of "Grand Hotel" deserves praise of a more than superlative order. He has used every artifice known to cinematography in so doing—yet without conveying the impression of artifice at any time.[7]

We might call this strategy the "gradual shift." Following the principle of differentiation, a cinematographer should vary the style over the course of a film. However, those variations should be subtle enough to maintain the principle of invisibility.

Just before *Grand Hotel*'s climactic murder scene, Preysing (Wallace Beery) is preparing to spend the night with Flaemmchen (Joan Crawford). The mood mixes comedy and drama: Preysing's humorous crassness makes Flaemmchen's difficult choice seem all the more unfortunate. The mood shifts to melodrama when Preysing steps into the next room and finds the Baron (John Barrymore) trying to steal his pocketbook. The Baron returns the pocketbook, but an enraged Preysing kills him anyway. Daniels handles this shift in mood with exceptional skill. The Preysing-Flaemmchen scene is played in a moderately high key. The presence of Venetian blinds is established, but Daniels refrains from using a potentially melodramatic shadow pattern here. When Preysing confronts the Baron, however, Daniels casts the blinds' pattern over Preysing's face, even though the blinds themselves are much less prominent in the shot. The key is darker, but it is not so dark as to lose detail. Daniels saves his darkest shadows for the murder itself. As Preysing realizes what he has done, half of his face is in shadow, and most of the set is cast into blackness. By following the convention of the gradual shift, Daniels expresses the changing mood of the story without letting the changing style become a distraction. As the drama ratchets up, Daniels employs a darker palette of shadows.

◉ ◉ ◉

As we have seen, the cinematographer had at hand a variety of ways to solve the problem of conflicting conventions: the sacrifice, the compromise, selective emphasis, and the gradual shift. The cinematographer had to think about a wide range of factors, and come up with a complex solution that could do justice to as many of those factors as possible. The emphasis might shift toward glamour at one moment, and then shift toward composition in the next. Effects might play a minor role for the bulk of the film, only to emerge more forcefully at the end, adding atmosphere to a crucial scene. This helps to explain why the style looks like a "neutral" style at a casual glance. With so many different conventions pulling in so many different directions, the style favors moderate solutions that fulfill a variety of aesthetic mandates. The dark shots never get too dark, because pools of blackness would conflict with the need for glamour and narrational clarity. The bright shots never get too bright, because excessive illumination would conflict with the need for atmosphere and modeling. This is not a neutral style—it is just remarkably subtle.

Situating this problem of conflicting conventions within an institutional context makes it more concrete. The unstated policy of the American Society of Cinematographers encouraged some of its members to put additional weight on some functions at the expense of others. For more than two decades the ASC had exerted great effort to create a public identity for the cinematographer as an artist. This institutional agenda favored aesthetic techniques like mood lighting and effect-lighting. However, cinematographers did not work for the ASC—they worked for the studios. These studios generally wanted cinematographers to place the most emphasis on a different set of techniques: the techniques of glamour lighting. As Irving Thalberg reportedly said, "I don't care what lighting scheme you use. My actors have got to look great, and actresses have to look beautiful. That's where the money is, right?"[8] This conflict between the aesthetic idealism of the ASC and the star-oriented strategy of the studios was one of the defining conflicts of Hollywood cinematography.

Discussing the conflict between the ASC and the studios should provide us with a new way of thinking about the classical Hollywood style. Historians who think of the Hollywood style as a "classical" style are happy to acknowledge that there were cracks in the system: the dominant norms called for unobtrusive storytelling, but there were always exceptional filmmakers who managed to subvert those norms with moments of stylistic spectacle. This is not the argument I am making; if anything, the reverse is the case. It was the ASC that encouraged cinematographers in this period to adhere to the norms of unobtrusiveness, because it was the aesthetically oriented ASC that encouraged cinematographers to think of themselves as storytellers. The ASC promoted conventions like mood lighting and effect-lighting precisely because these were the conventions that turned cinematographers into storytellers. By contrast, it was the powerful studios that encouraged cinematographers to drop their storytelling conventions when they got in the way of glamorous photography. For instance, James Wong Howe once shot an unglamorous close-up of Myrna Loy for a scene in which her character was supposed to look at a mirror and comment on how bad she looked. Howe assumed that he was simply serving the story, but studio boss Eddie Mannix rejected his opinion, demanding a more glamorous reshoot.[9] Howe's rebellion here was not the choice to favor spectacle over story, but the opposite: he wanted to serve the story, and the studio was demanding more spectacle. In other words, we should not think of cinematographers as pictorialist aesthetes who were forced to submit to the storytelling norms. Rather, we should think of cinematographers struggling to tell the story, while simultaneously fulfilling the mandate to produce glamour.

While at MGM Howe had shot some films for W. S. Van Dyke (who was married to Mannix's sister). Van Dyke was famous for being the fastest director in Hollywood, but even "One-Shot Woody" would alter his working habits when it came to glamour. According to cinematographer Harold Rosson:

> His thinking, as I recall it, was to emphasize certain things in a picture. In other words, the girl in his picture had to look beautiful. So Van would permit you as a cameraman as much time as needed, within reason, to get a good result of the girl. If I was going to make a photograph of a piece of newspaper that had to be lying on that table, he wanted me to photograph that newspaper in one second, or half a second if possible, but I could take hours with the girl.[10]

This story suggests that glamour was not just a subordinate principle used only when it could help the narrative. Glamour was a dominant principle, sharing equal footing with narrative clarity. Storytelling was supposed to be efficient, but glamour photography was supposed to be spectacular—particularly when female stars were concerned.

Of course, there are certainly examples of cinematographers who were willing to sacrifice glamour to enhance the story. We can even find some examples at MGM. Oliver Marsh, who used a top-light to mark Jimmy Stewart's insanity in *After the Thin Man* (fig. 7.2), also employed a top-light for a dramatic moment in the fallen-woman film *Faithless* (Henry Beaumont, 1932). In the film, the Depression causes Carol (Tallullah Bankhead) to turn to prostitution to save her sick husband. When she is at her lowest point, Marsh places her under a harsh top-light, stripping the glamour away from a destitute character who has lost everything.

In *The Most Dangerous Game* (1932), cinematographer Henry Gerrard makes the opposite choice, opting to preserve the glamour of Joel McCrea in a scene that normally would have called for a dramatic effect. Figures 7.9 and 7.10 show two men who are supposedly standing next to each other, but they look like they are in completely different films. In this case, the problem involves balancing at least four separate needs: the need to create a mood of melodramatic menace around the villain, the need to keep Joel McCrea looking good, the need to make all the lighting appear motivated, and the need to prevent style from becoming a distraction. On my scorecard, Gerrard manages to get two and a half out of four. He emphasizes Leslie Banks's menace by using extreme contrast and a low-placed key-light. He flatters McCrea's handsome looks by keying from the side, with a moderate amount of fill. The light for Banks follows the conventions of effect-lighting, since it has been motivated by a source shown in a previous shot. However, the light for McCrea violates the same

Figure 7.9 The villain is lit from below in *The Most Dangerous Game* (1932).

Figure 7.10 In the same scene, Joel McCrea is lit for glamour.

principle of motivation, since it is clearly inconsistent with the light for Banks. This, in turn, causes Gerrard to break the rule that style should not take attention away from the story. Does this mean that Gerrard has employed a daringly unclassical lighting scheme? Not necessarily. He has sacrificed the ideal of invisibility, but he has accomplished some equally important goals, expressing the mood of the story while preserving the glamorous persona of the star.

In his Academy Award–winning work on *A Farewell to Arms* (1932), cinematographer Charles Lang crafts an even more efficient solution. Figures 7.11 and 7.12 demonstrate the standard figure-lighting techniques that Lang uses for the bulk of the film. Although Helen Hayes and Gary Cooper are in the same space, Lang handles their close-ups differently. Specifically, there are differences in the direction of the key-light and in the use of lens diffusion. Hayes receives a frontal key, from a light placed almost directly above the camera. (This light casts the shadow under Hayes's chin.) Because of its frontal placement, it leaves no strong shadows on Hayes's face; even her nose casts no visible shadows. This has the effect of turning her face into a flat, smooth plane against which the eyes and lips stand out. By contrast, Cooper's key comes from a front-cross position, lighting the far side of his face while keeping the camera side in shadow. The shadow receives some fill, but not so much as to cancel out the difference between the two sides of his face. Cooper receives more "modeling," and his face appears remarkably three-dimensional in comparison to Hayes's "flatter" treatment; for instance, the shape of his nose and the area around his mouth receive more detailed handling. At the same time, Hayes's shot contains more lens diffusion, which further "softens" her features. In short, Lang shows that he is in full command of the gender-specific figure-lighting conventions outlined in chapters 2 and 6.

We get a better sense of Lang's accomplishment in figures 7.13 and 7.14. In this scene, a drunken Cooper examines the shape of Hayes's foot, while they hide in an alley during a bombing raid. The time (night), the

Figure 7.11 Frontal lighting for Helen Hayes in *A Farewell to Arms* (1932).

Figure 7.12 In the same scene, Gary Cooper is lit from a front-cross position.

location (an alley), and the type of scene (battle scene) all call for a darker tonality than was used in the previous example. The remarkable thing is that Lang accommodates these competing functional needs without abandoning his carefully differentiated figure-lighting strategy. The placement of Cooper's key is almost identical, but his backlight has been eliminated, and he receives little or no fill light. The result is a high-contrast image, with a dark overall tonality. This dark tonality fulfills the above-mentioned functional needs, but the lighting strategy still manages to follow the logic of emphasis and expression: the placement emphasizes the character-defining features of his face, while the "strong" contrasts express his "masculinity" via associations. Meanwhile, Hayes's key-light is still essentially frontal, though it does cast a bit more shadow on the far side of her face. The key is at a lower level of brightness, relative to the exposure. This allows Lang to achieve the necessary dark tonality, while creating a "smooth," low-contrast image. Heavy lens diffusion smooths the image even more. Again, the strategy follows the dual logic of emphasis and expression: the frontal placement deemphasizes facial lines, while the "gentle" gradations in tonality express the idea of "femininity." Presumably, it was multifunctional solutions such as these that helped Lang win the Academy Award for his work on this film.

Charles Clarke has explained some of the specific problems of producing glamour in exterior shots. Clarke was an expert in exteriors, and he often used a red filter to darken the sky, making the white clouds stand out. This had the advantage of producing a dynamic pictorial effect, but it also created some new problems. Clarke writes:

> We must remember that in the movies we never hold these pictorial scenes very long on the screen, for we must get on with the story. If we were using the very red #29 filter on every scene of this sequence, we might get poor reproduction if we filmed the leading lady up close with the #29 filter. It would turn her face to a chalky white and the high contrast would be ruinous to her beauty.
>
> After the long shots, we would change to a less severe correcting filter for the close-ups. . . .
>
> If there were no leading lady in this series of scenes, only men without makeup, then we would not be so concerned about the facial reproduction. The men, without make-up, look better with high contrast and over-correction, for it gives them ruggedness and strength.[11]

Clarke recommends the gradual shift strategy, emphasizing the contrast in the wide shots, when the pictorial function is most critical, then subtly

Figure 7.13 A low-key, low-contrast image of Hayes.

Figure 7.14 A low-key, high-contrast image of Cooper.

reducing the contrast for closer shots, when glamour becomes a more important function. Following the logic of the figure-lighting conventions, this strategy would be applied differently for men and women.

Some cinematographers opted to use glamour lighting selectively, saving it for the most romantic scenes. In an oral history, told to Alain Silver, James Wong Howe tells a story about lighting Hedy Lamarr in *Algiers* (1938):

> In photography there are certain moments you build up to in lighting, the same as with music or drama. After reading the script I felt that when they meet, their love scene, there was, I thought, a moment where she should look really the most beautiful. And in all those close-ups I took particular care and attention, care in lighting and the camera angle. We got huge close-ups: the mouth, the eyes, the throat and so on. And we must have spent two or three days in photographing these close-ups for this love sequence; and she looked really beautiful. After she saw the rushes, she said, "Well, you know, I would like to look like that way throughout the whole picture. So why don't we make

all the others over again?" I told her that I thought it would be the wrong photography, the wrong approach, because then when we come to this love scene
we had no place to go.[12]

They eventually agreed on this approach. By refusing to overtly glamorize
the entire film, Howe created a special moment where glamour and
mood could work together.

While these are complex but effective solutions to the problem of conflicting conventions, some cinematographers wondered why they had to
tackle this problem at all. Why couldn't cinematographers ignore the
mandate to produce glamour when the story called for an unglamorous
style? Here Victor Milner was a crucial voice in the debate. Milner, an
Oscar-winning Paramount cinematographer, wrote several articles on
this topic for the ASC throughout the thirties. Milner complained repeatedly that the mandate to produce glamour had become oppressive. For
instance, in his 1930 essay "Painting with Light," Milner writes:

> Regardless of what the story may call for, the star must at all times be so pho-
> tographed as to be the outstanding feature of the scene. This is a hardship
> alike upon the director and the cinematographer, for there are times when
> even an extra may be of more dramatic or artistic importance than this famous
> profile, or that famous pair of nether extremities. In addition, a star must al-
> ways be so photographed as to be perpetually at his or her best. No matter
> whether the action be taking place in a dungeon or a ball-room, the star must
> be kept beautiful, with never a suspicion of a shadow upon the famous face or
> form, and, regardless of the setting, a beautifying halo of back-lighting follow-
> ing her around the set. That this is generally illogical and inartistic carries lit-
> tle weight with people whose inflated egos are backed up with contractural
> [sic] requirements for perpetual photographic partiality.[13]

The alliterative prose suggests to me that *AC* editor Hal Hall had a hand
in crafting Milner's essay. Still, it is certainly plausible to suppose that the
angry tone originated with Milner—particularly when we consider the
fact that Milner returned to these themes again and again.

The cinematographer was an artist in large part because he was a storyteller, but glamour lighting was a threat to storytelling on several fronts.
The cinematographer could not always direct attention to the most salient
story point, because attention had to be kept on the star at all times. The
cinematographer could not always light for mood, because tonality, contrast, and diffusion might be determined by glamour considerations. The

cinematographer could not always denote time and space realistically, because the star had to be lit a certain way, at all times, in all spaces. Glamour lighting undercut the aesthetic rationale behind most of the cinematographers' other conventions, including those of composition, genre, and effect-lighting. Glamour lighting did not even support the ideals of figure-lighting, since the most artistic figure-lighting was supposed to reveal the depths of a fictional character, not to idealize the surface of a studio star.

The fact that Milner endorses storytelling as his highest ideal does not mean that he is simply submitting to the classical ideal of clear narration. First, we should remember that Milner defined storytelling in an expansive way, placing special emphasis on the use of expressive techniques like mood lighting. Second, when it comes to lighting, the discourse of the ASC was not a response to preexisting ways of thinking about light—rather, it helped to constitute the ways that Hollywood practitioners thought about light. Third, it seems odd to say that the ASC was submitting to a dominant norm when they often had to struggle against the powerful studio bosses to protect the ideal of storytelling against the threat posed by the studio demand for glamour. From the point of view of the ASC, the endorsement of storytelling as the preeminent ideal was a kind of aesthetic protest against the philistinism of the glamour machine.

Turning to specific cases, we find that the glamour problem could be solved in various ways. In a 1935 article, Milner explains the different strategies he used to light *Cleopatra* (1934) and *The Crusades* (1935), both directed by Cecil B. DeMille. Regarding the latter, he writes:

> The women in the story demand the softest, most delicate lighting possible to fittingly express the Medieval ideal of womanhood. The "emancipated woman" of today was nearly ten centuries in the future, and the twelfth century Chivalric concept of womanhood was as of something fragile, ethereally lovely; almost too wonderful for the rough men of the period to even touch. This concept can be carried out in the lighting, soft, delicate lighting, possibly higher keyed and attuned to a greater degree of photographic diffusion.[14]

This is a complicated argument. While Milner reinforces the logic behind the differentiating conventions—calling for soft, low-contrast lighting for the film's female stars—he also suggests that this treatment is more appropriate for a film set in the past, precisely because the basic oppositions had been blurred in recent years. The implication is that films about modern women should avoid the softening strategies that were developed during the silent period.

Milner applies the same double-edged argument to his discussion of his lighting strategies for the film's male stars:

> Both characters represent the flower of hardy, virile manhood, and tower above their fellows in body and in character. This gives an unusual opportunity for psychologico-dramatic cinematography. Strength and virility must predominate in the lighting of these men, who should be shown in strong relief against the pictorial, but less commanding background of the period and of their followers. To this end, I hope to depart from the conventional lighting technique; instead of avoiding contrasts and strong shadows, for instance, I plan to heighten them to express visually the strength of the two men. Using as little as possible of conventional flat front-lightings and outlining back-light, I plan to simplify my personal lightings, throwing features into strong relief, modelled by heavy shadows that emphasize the note of ruggedness.[15]

By suggesting that this technique was tailored to the needs of a period film, Milner indicates that Hollywood's romantic male leads were generally not rugged enough to merit the hard side-lights and weak fill lights of the "virile" treatment. Again, he reinforces the logic behind the differentiating conventions while simultaneously questioning their contemporary relevance.

Milner is certainly exaggerating when he implies that his techniques were daringly original. Well into the thirties, we can find the sentimental ideal of womanhood alive and well in the romantic drama, and we can find the tough guy brandishing his virility in the crime film. Because of this we can find plenty of modern-day stories with soft, smooth lighting for women, and hard, contrasty lighting for men. However, Milner is right to suggest that many cinematographers chose the simplest strategy, using a simple three-point setup for men and women alike.

Milner's discussion of *Cleopatra* is even more complex. Here he proposes a strategy that prefigures some important ideas in contemporary film theory:

> The story of "Cleopatra," for example, might have been interpreted in innumerable different ways by different Directors: one might have chosen to concentrate upon the purely emotional phases of the character, while another might have made Cleopatra herself of secondary importance to Caesar and

Antony. In either case, the photographic treatment—especially the style of lighting—would have to be basically different. Cecil DeMille, in his recent production of the story, which I photographed, emphasized yet another angle: the fact that Cleopatra was essentially a "showman," who instinctively dramatized her every surrounding, and amazed the luxurious Roman world with her calculatedly lavish display of Egypt's incredible wealth. Thus in this production, the photographic keynote was richness and in every scene—even the most dramatic—the lighting was kept richly brilliant, to keep the audience subtly aware of the splendor of the settings and costumes. Similarly, Cleopatra's entertainments of Caesar and Antony were lit and photographed primarily as sensuous spectacles, the high-lights of her exhibitionistic nature.[16]

When we compare this passage to the film itself, we find some evidence supporting this account—as well as some that calls for additional nuance. First, it is true that Cleopatra's scenes are consistently bright, allowing us to marvel at the enormous sets as they sparkle with Milner's highlights. Figure 7.15 shows the moment when Cleopatra first learns about Caesar's death. This is a serious dramatic scene, and Milner might have used some shadows to darken the mood. Instead, he increases her glamour by giving her a soft, frontal key and a strong backlight. In some scenes, Antony and Caesar receive similarly glamorous treatment. However, Milner varies the pattern by employing genre/scene conventions at crucial moments. When Caesar is killed, Milner uses the low-placed key-lights of the melodrama. When Antony is facing death at the end of the film, Milner uses the strong shadows of the serious drama—as in figure 7.16, where Milner lights Antony with minimal fill and a hard side-light.

We could explain this difference by pointing to the gender-based figure-lighting rules, noting that the lighting smooths the features of Cleopatra, while modeling the features of Antony. However, this explanation does not go far enough. Here, we must note that Milner has constructed a gender-based difference on a new level. We might call this level the "range of variation." It is not simply that he lights Cleopatra differently from how he lights Antony. It is that Antony's lighting is allowed to *vary* over a greater *range* than Cleopatra's lighting. While Cleopatra's lighting is consistently brilliant, as Milner suggests, Antony's lighting ranges from brilliant to somber.

This may seem like a small difference, but it gains added significance when we consider Milner's primary reason for restricting Cleopatra's range of variation: her penchant for exhibitionistic spectacle is seen as the key to her character. Ever since the publication of Laura Mulvey's

Figure 7.15 Claudette Colbert is lit for glamour throughout *Cleopatra* (1934).

Figure 7.16 *Cleopatra*: Somber tonalities for the scene of Marc Antony's death.

landmark essay "Visual Pleasure and Narrative Cinema," film scholars have examined the ways Hollywood films situate women within a position of "to-be-looked-at-ness." Mulvey's analysis examines the opposition between narrative and spectacle. Whereas male figures typically take active roles in a narrative, female images are offered as nonnarrative spectacle, presented to the spectator for fetishistic enjoyment.[17] This example from *Cleopatra* both complicates and confirms Mulvey's point. On the one hand, it suggests that the opposition between narrative and spectacle can be broken down. Milner uses a spectacular style for Cleopatra precisely because such a style is well suited to the narrative. The presentation of DeMille's trademark kitsch is justified by the need to characterize Cleopatra herself as a "showman"—an active woman who becomes powerful through the strategic use of exhibitionism.

On the other hand, the integration with the narrative is not complete. In order to achieve maximum integration, a cinematographer would vary the style over the course of a film, perhaps adjusting for a shift in location or a shift in mood. While Cleopatra's consistent brilliance is given ample narrative justification by her use of spectacle as a strategy, the very consistency of that brilliance limits the degree to which Milner can make use of all the basic tools that he would normally rely on to help tell the story. In other words, because he is committed to making the point about Cleopatra's exhibitionism over and over again, Milner is willing to sacrifice all the subtle modulations of style that are typically used to express a character's emotional arc. The style expresses the emotions of Antony, but not those of Cleopatra. In other words, the lighting characterizes him with greater complexity.

This concept of "range of variation" can be a very useful tool in the analysis of Hollywood cinematography. One cinematographer might employ a large range of variation for both men and women, and another cinematographer might employ a limited range for both. However, a surprising number of major Hollywood cinematographers chose a third option: employing a large range for men and a small range for women. This was a very effective compromise, satisfying the ASC's desire to see cinematographers apply the right-mood-for-the-story theory, while satisfying the studios' desire to see female stars represented glamorously at all times.

For example, Joseph Ruttenberg employs this compromise in Fritz Lang's *Fury* (1936). Most of the film is shot in a sharp, gray style, in the realist manner. In the first part of the film, Sylvia Sidney and Spencer Tracy are both lit to look unpretentiously attractive. As the jail is burned down, Ruttenberg switches to the conventions of melodrama: the palette is darker, and the contrasts are stronger. Some members of the lynch mob are even lit with low-placed keys. Tracy's lighting also shifts to a darker register, but Sidney's lighting is curiously unaffected (fig. 7.17). She receives less fill light than before, but her key-light still has the same frontal placement that was standard for images of women. Later, when Joe (Tracy), thought to be dead, makes a sudden reappearance, Ruttenberg uses an unusually dim key-light to heighten the air of mystery (fig. 7.18). Sidney, however, is presented in the same way as before, with a bright frontal key highlighting the emotional expressions on her face. In short, the most important difference between the handling of Tracy and the handling of Sidney is not found at the level of the device; it is found at a more abstract level, in the style's range of variation.

Figure-lighting conventions and genre/scene conventions can often converge because they both rely on an association-based logic that is structured by the larger culture's ideas about gender and emotional experience. These conventions may also be influenced by the culture's ideas

Figure 7.17 Bright frontal light-ing for Sylvia Sidney in *Fury* (1936).

Figure 7.18 *Fury*: Spencer Tracy in shadow.

about gender and emotional expression. In *Fury*, when Katherine (Sidney) watches the jail burn down, the frontal lighting makes her emotional expressions easy to read. Her face is nothing if not visible, and we see the emotions on her face directly. When Joe returns from the dead, he does his best to keep his rage bottled up inside; the lighting comes to his aid by putting his face in shadow. Is it going too far to say that the style does some of the expressive work for him?

The relationship between gender and the expression of emotion is one of the explicit themes in Howard Hawks's *Only Angels Have Wings* (1939, shot by Joseph Walker). In typically Hawksian fashion, the story concerns an all-male cadre of flyers. When one of the pilots dies, the flyers refuse to mourn him. Instead, the lead flyer Geoff (Cary Grant) blames the pilot: "He just wasn't good enough." In equally typical fashion, a Hawksian woman (Bonnie, played by Jean Arthur) is allowed admittance into the group only when she demonstrates her ability to follow their rules.[18] However, in the end, Geoff breaks the stoic masculine code, crying when his best friend the Kid (Thomas Mitchell) perishes in a plane crash.

Joseph Walker, the master of selective emphasis, shot the film. He refuses to use the somber style for every serious scene. The introduction of
the unredeemed Kilgannon (Richard Barthelmess), for instance, is certainly a dramatic scene, but it is handled in a high key. Walker reserves
his deepest shadows for the death scenes: the death of Joe (Noah Beery),
the death of the Kid, and the near-death of Geoff. These scenes highlight
Walker's ability to compromise. His basic lighting strategy for Bonnie
relies on the standard conventions for lighting women: a bright frontal
key, generous amounts of fill, and strong backlights to emphasize Jean
Arthur's blonde hair. While Geoff and the Kid are swallowed by dark
shadows during Joe's death scene, Bonnie, standing a few feet away, receives more light. She even receives more light than the two men standing next to her. Later, another man consoles Bonnie after she has been
kicked out of the bar by Geoff. The man's lighting is somber, but Bonnie's
face is lit up with a strong frontal key, while her hair glows with a blazing
backlight (figs. 7.19 and 7.20). The pattern is clear: the men, who hold
their emotions in check, are lit with somber tonalities during the tragic

Figure 7.19 In *Only Angels Have Wings* (1939), the men are often lit with more shadows.

Figure 7.20 In the same scene, Jean Arthur receives more light, even though she is facing away from the light source.

scenes. Bonnie's emotionally expressive face, by contrast, is—and must be—consistently visible.

The most remarkable example of this technique appears in the scene after Geoff's brush with death. The Kid consoles Bonnie, who is still struggling with Geoff's insistence on bottled emotional reserve. A carefully motivated Venetian blind pattern provides a somber mood as it casts a shadow over almost everything in the room. The one exception is Bonnie's face, which receives a curiously unmotivated light on her face, highlighting her emotional expressions (fig. 7.21). There is a certain division of labor in this scene. The woman's performance is more expressive, and the cinematographer can highlight the emotion best by directing our attention to her brightly lit face. While it would be an overstatement to say that Thomas Mitchell's subtle performance is unexpressive, it does seem fair to say that the somber style manages to do a lot of the expressive work for him.

We can partly explain this pattern by appealing to gender-based figure-lighting conventions. Women are lit for glamour by lighting up their faces until their features are smooth. In his autobiography (co-authored with Juanita Walker), Joseph Walker described "Harry Cohn's edict: these gals must look glamorous regardless of the role."[19] Regardless of the drama that Jean Arthur's Bonnie must endure, she must always be lit for glamour. The fact that this rule applied to women in particular is underscored by the title of the Walkers' book, *The Light on Her Face*.

More important, we should remember that cultural ideals about stoic men and expressive women shape the ways that fictional characters reveal their emotional states. This is true of most Hollywood films, but it is particularly true of *Only Angels Have Wings*, which frames the relation-

Figure 7.21 Shadows are cast throughout the composition, but not on Jean Arthur's face.

ship between gender and emotion as an explicit theme. A comparatively inexpressive character (in the world of *Angels*, Geoff) has more to gain from the extra expressivity that the full range of genre conventions can provide. By varying across a wide range, the mood conventions can enhance the emotional arc of a character who is doing everything to keep his emotions bottled up inside. Meanwhile, an emotionally expressive character (in the world of *Angels*, Bonnie) needs to be lit to make her facial expressions visible. The desire to keep the woman's emotions visible produces a preference for consistently bright tonalities; it requires "the light on her face."

The end of the film provides the exception that proves the rule. After the Kid dies, Bonnie visits Geoff and sees him crying for the first time. For this rare moment in the film when Cary Grant's face becomes the site of uninhibited emotional expression, Walker hits his face with all the light he can.

Only Angels Have Wings represents one solution among many to the problem of conflicting conventions. Whereas some cinematographers are willing to sacrifice glamour on occasion whether the subject is a man or a woman, and others devise creative compromises in order to fulfill multiple mandates at once, there are still others who solve the problem by restricting the lighting options—often, the lighting options for women—to a narrow range of variation, on the brighter side of the scale. This keeps their expressions visible, while fulfilling the studio's mandate to make them look glamorous at all times.

These examples serve to remind us yet again that the style of Hollywood lighting was not a single list of rules that could be applied unthinkingly to any situation. There was a wide range of strategies, producing both convergence and conflict. This is what we should expect when we are dealing with a competitive group of filmmakers who share similar interests—not a harmonious set of perfectly ordered principles, nor a battleground of contested strategies, but something in between. When we take a global view of the thirties, we see broad areas of agreement. Most cinematographers were familiar with the sets of conventions, and endorsed the aesthetic justifications behind those conventions. When we make a more localized examination of individual films, we see that each film presented its own array of problems, and that each cinematographer could prioritize the conventions differently to achieve the desired balance.

All of this emphasis on difference runs the risk of making the category of classical Hollywood lighting too broad, able to incorporate almost any deviation from the norm. In order to mitigate against this tendency, I want to propose one more distinction, a distinction that will allow us to classify cinematographers on the basis of their adherence to the ideal of balance. Some cinematographers operate at the margins of the classical style, favoring a mannered approach that sacrifices certain functions in order to intensify others. Other cinematographers work at the center of the classical style, always searching for an optimum balance. As a paradigm for this distinction, we can consider a disagreement between Lee Garmes and William Daniels. In an oft-quoted passage, cinematographer Lee Garmes describes how he created his most famous image:

> Unfortunately I didn't have sufficient time to make tests of Marlene Dietrich; I had seen *The Blue Angel*, and, based on that, I lit her with a sidelight, a half-tone, so that one half of her face was bright and the other half was in shadow. I looked at the first day's work and I thought, "My God, I can't do this, it's exactly what Bill Daniels is doing with Garbo." We couldn't, of course, have two Garbos! So, without saying anything to Jo [von Sternberg], I changed to the north-light effect. . . . The Dietrich face was my creation.[20]

Significantly, Garmes does not say anything about selecting a light that is suitable for the story. Instead, Garmes's guiding ideals appear to be glamour and pictorial quality. He even adds a note of artistic refinement by referring to his technique (inaccurately) as the "north-light" effect.[21]

Garmes was trying to pay Daniels a compliment, but Daniels probably would not have taken it. In his own interview with Higham, Daniels says:

> Unlike, say, Lee Garmes, I vary my work considerably according to the story. Even my lighting of Garbo varied from picture to picture. There wasn't one Garbo face in the sense that there was a Dietrich face. . . .
>
>   I didn't create a "Garbo face." I just did portraits of her I would have done for any star. My lighting of her was determined by the requirements of a scene. I didn't, as some say I did, keep one side of the face light and the other dark.[22]

Daniels bristles at the suggestion that he cared only about Garbo's glamour. While he admits that he created "portraits" of Garbo, he insists that he adapted those portraits to suit the needs of storytelling. In so doing, he adheres to the norms of classical Hollywood lighting.

In comparing these two passages, it is tempting to say that Garmes is a confident artist, while the more classical Daniels has accepted his role as a humble craftsman. However, this account does not capture the tone of the comparison. They are both confident artists; they just define art differently. For Garmes, the cinematographer creates beautiful images. For Daniels, the cinematographer searches for an ideal synthesis, serving the story and the star at the same time.

My proposal is that the conceptual distinction between Daniels and Garmes is an example of a more general distinction—a distinction that will allow us to make sense of the wide variety of stylistic options that Hollywood cinematographers explored during the studio period. On the one hand, some cinematographers are committed to an art of balance, attempting to fulfill as many different functions as possible with their stylistic decisions. This category would include such masters as William Daniels, Charles Lang, Victor Milner, and Joseph Walker. Because balance is often seen as a classical ideal, we might call these cinematographers the "classicists." The classicists try to avoid sacrificing one ideal for another, preferring more nuanced solutions like the compromise, the gradual shift, and selective emphasis.

On the other hand, some cinematographers opted to explore the possibilities offered by one function, even if it meant sacrificing some of the other functions. For instance, a cinematographer might consistently sacrifice realism in favor of glamour, or sacrifice narrative clarity in favor of mood. This category of cinematographers, including Lee Garmes, Gregg Toland, John Alton, and Leon Shamroy, can be called the "mannerists." Of course, these cinematographers had very different styles, so it might be useful to think of this category as a general term comprising various subcategories, such as hyper-realists (who push the limits of the realist ideal), expressivists (who create expressive effects at the cost of glamour and clarity), and pictorialists (who foreground pictorial values like beauty and innovation).

One obvious source for this proposal is the work of Fabrice Revault d'Allonnes, mentioned earlier, in chapter 3, a French historian of lighting whose analysis of lighting is based on a three-part distinction between classical, baroque, and modern lighting.[23] However, the proposal also has a less obvious source: the essay "Norm and Form," by the art historian Ernst Gombrich.[24] In this essay Gombrich opposed a tradition of art historians who would see a Mannerist or Baroque painting as a manifestation of anticlassicism. For Gombrich, an emphasis on what the painter is *not* doing obscures the fact that the typical Mannerist or Baroque artist has a great deal in common with the Renaissance artist. As an alternative, Gombrich suggests that the history of Renaissance, Mannerism, and Baroque art is partly shaped by the conflict between two principles: the principle of order and the principle of fidelity to nature. These two functional aims are not mutually exclusive, but they can be mutually limiting. A highly realistic picture might appear overly chaotic; an overly ordered composition might violate conventions of realism. Gombrich suggests that the paintings of the High Renaissance were influential in part because they seemed to offer an ideal synthesis, keeping sacrifices to a minimum on both sides. In comparison to this ideal, a Pontormo or a Caravaggio might initially appear anticlassical, but Gombrich suggests that the real difference is more subtle: it may be that the Mannerist or Baroque painter is simply focusing his or her energies on the exploration of one of the two principles, thereby sacrificing more of the other principle than a more classical painter would be willing to do. The difference is real, but in comparison to a genuinely anticlassical artist (say, Duchamp), the similarities are equally important. The High Renaissance artist, the Mannerist artist, and the Baroque artist all value order and fidelity; they just prioritize them differently.

Whether Gombrich's account of this episode from art history is correct or not, his essay offers a model for the distinctions I am proposing. All the cinematographers of the ASC valued the same set of ideals: storytelling, realism, glamour, and pictorial quality. Based on the nature of the task at hand, there was always the potential for conflict among these functions within a film. Indeed, there was always the potential for conflict within functions—for instance, there might be tension between the storytelling functions of mood and clarity, or between the realism of effect-lighting and the realism of illusionism. Amid all these potential conflicts, the classicists aimed for an ideal synthesis, achieving a minimum of sacrifice on all sides. The mannerists had the same set of values, but they had a different set of priorities. Rather than search for an ideal synthesis of all the functions, they explored the artistic possibilities offered by one or two distinct functions. Both kinds of cinematography are artistic, but in different ways. A mannerist cinematographer might practice an art of experimentation, trying out new ways of fulfilling a single ideal. A classicist practices an art of balance, allowing the ideals of storytelling, realism, glamour, and pictorial quality to dovetail. The two approaches can produce results that are equally accomplished, and equally deserving of our attention.

Not surprisingly, my account of the classicists draws on the tradition of Bordwell, Staiger, and Thompson, who have argued persuasively that the classical Hollywood cinema was a stable system, able to incorporate a wide variety of techniques.[25] Indeed, Bordwell himself cites Gombrich's essay as an important model. Still, my argument departs from theirs in one important way. Bordwell, Staiger, and Thompson argue that the Hollywood style served a wide range of functions, but they also contend that one particular function has dominant status: the function of clear narration. I would suggest that the concept of the "dominant" is not always the most helpful framework. We should invoke this notion only in a case where there is a struggle between competing ideals, and one must prevail.[26] Classicists did not always seek to subordinate one function to another; rather, they often sought compromise solutions that would satisfy as many different ideals as possible. As evidence for their account, Bordwell, Staiger, and Thompson would point out that it is very hard to find stylistic techniques that fail to contribute to the story. I agree, but I would add that this evidence is compatible with another hypothesis. Hollywood cinematographers were indeed reluctant to sacrifice the story, but they were often reluctant as well to sacrifice other ideals like glamour and the illusion of roundness. By finding judicious compromises and employing gradual shifts in technique, they crafted a style that would suit the ever-present and ever-important demands of the story, the studio, and the star. For the classicists, multifunctionalism was the guiding ideal serving a set of rival constituencies.

In his films for Josef von Sternberg, Lee Garmes is a consummate mannerist, pushing the limits of pictorialism. In *Morocco* (1930), one shot shows us Tom Brown (Gary Cooper) knocking on a door (fig. 7.22). Garmes and von Sternberg take the opportunity to craft an image with a bewildering variety of pictorial patterns: notice the vertical lines of the stairway on the right, the curved diagonals of the large leaf in the foreground, the crisscross shadow pattern on the center wall, the horizontal lines of the door itself, and the sinuous line of the column on the left. It is a dazzling image, but we should be cognizant of what Garmes and von Sternberg were willing to forgo. First, they sacrifice the illusion of roundness: like a modernist painting, the image is a series of flat planes and crossing lines. Second, they sacrifice the advantages that normally come with effect-lighting: the cast shadow on the wall does not really help us understand the space, and it is not an obvious indicator of the time of day. Third, and most important, they sacrifice narrative clarity: Brown's face is in shadow, and his figure gets lost in the intricacies of the forms. A more classical cinematographer would not necessarily have eliminated all of these patterns. Rather, he would have rearranged the composition, allowing Brown to dominate a pictorially pleasing frame.

Figure 7.22 A nearly abstract composition in Josef von Sternberg's *Morocco* (1930).

Significantly, Garmes is not a Hollywood outsider. In fact, he frequently endorses classical ideals; he just interprets them in a way that allows him to practice a particular brand of pictorialism. In a 1938 article he writes, "I would be in the end a bad photographer if I created photographic gems which shone so brightly that they dazzled the spectator and diverted his interest from the purpose of the scene as a whole." At first this rings as a clear endorsement of the invisible ideal, but Garmes goes on to find some loopholes in the rule:

> The exception to that, of course, is when there comes a scene in which the whole point is the photography. But even then the photographer's highest purpose is achieved in making that scene exactly right in its context, which is in accordance with the general rule I have been expounding. Tricks executed by the photographer alone nearly always upset the balance of a film. There may come a time when the photographer is called upon to bring forth a photographic *tour de force* to strengthen a dramatically weak scene, or to introduce novelty in what would otherwise run the risk of being commonplace.[27]

The first loophole is the fact that certain stories demand beautiful photography. *Morocco* would fall into this category, since it was a romantic drama set in an exotic location. The second loophole presented is the fact that most films contain scenes that are, to put it bluntly, boring. A beautiful shot can maintain the audience's interest until the drama revives. Ironically, attracting the audience's attention to the photography for the short term is actually a strategy to keep the audience's attention on the story for the duration of the film.

Rather than thinking of Garmes and von Sternberg as forward-thinking ground-breakers, it might be useful to view them as filmmakers who never shed the ideals of the silent period, when cinematographers like John F. Seitz and Walter Lundin had proposed that pictorial beauty was perfectly compatible with Hollywood narrative. Early in his career, Garmes worked with John Leezer, the cinematographer who introduced the first soft-focus lens. Garmes had also worked with Rex Ingram, who actively encouraged him to imitate the style of Seitz.[28] This tradition encouraged Garmes to think of bold pictorialism as a viable option, even within a storytelling cinema.

Watching a von Sternberg/Garmes collaboration can be quite an experience: the key-light placement is daring, the range of contrasts is remarkable, and the vast array of shadow-patterns is astonishing. Watching a William Daniels films is less surprising, but I cannot help but think that Daniels is tackling a more difficult problem. While giving the star an eye-light, he must imitate sunlight, candlelight, or a lamplight. While emphasizing the star's hairstyle, he must emphasize the scene's primary plot point. While modeling the curve of a cheekbone, he must model the shape of the film's emotional curve.

The previous chapter provided some examples of Daniels's skills from *Anna Karenina*, such as his skillful handling of lighting in the two-shot (fig. 6.3), and his technique of putting a little extra light on the foreground (fig. 6.16). To this list we can add his selective use of expressive effects. Here Daniels follows the tradition of using the effect-lighting conventions to motivate the genre/scene conventions, always adjusting these conventions in subtle ways to accomplish additional functional goals. For instance, one scene shows Anna visiting her son's bedroom after she has spent the evening with Vronsky. To emphasize the serious mood, Daniels uses somber tonalities for the scene. These shadows are plausibly motivated by the presence of a sole light in the room—an unusual light that casts a striking pattern on the wall. The problem is that these dark tonalities might make it difficult for us to see the acting. This is, after all, a Garbo movie, and her acting ability is one of the film's primary attractions. To solve this problem, Daniels does what every great classicist would do—he cheats. To put it more politely, he devises an intelligent compromise. While keeping the general illumination low, Daniels puts extra light on the foreground, where the actors can perform. When Garbo enters the room, she passes through a pool of shadow, reinforcing the serious mood. But when she sits on the bed, a bright key-light gives her some unmotivated illumination (fig. 7.23). This scheme ensures that we will not miss a moment of Garbo's acting. The result is an elegant solution, combining the effect-lighting conventions, the genre/scene conventions, and the composition conventions to produce a multifunctional scene. Significantly, Daniels opts for a different solution during a later scene that also takes place in the son's bedroom. When Karenin visits his son, Daniels keys them from behind, darkening the overall mood while maintaining the consistency of the effect (fig. 7.24). There are several reasons for this shift. For instance, Basil Rathbone's acting was less of an attraction, and he did not require the glamorous treatment of a Garbo. Most important, however, the mood of the story has grown more tragic, justifying the use of a more somber style.

It may seem curious that some very dramatic scenes are lit with fairly light tonalities. For instance, Daniels uses a bright frontal key when Anna sees Vronsky for the final time. It might be pointed out that this is a daytime scene, but this explanation is not sufficient. Director Clarence

Figure 7.23 *Anna Karenina* (1935): Extra light on the foreground in a darkened room.

Figure 7.24 The same room with less light on the foreground.

Brown could have decided to stage this scene at night, or Daniels could have suggested daytime while lighting Garbo from the side. A more likely explanation is that Daniels is using selective emphasis, saving his shadows for a more important scene—namely, the emotionally somber scene when Anna commits suicide by throwing herself under the train. In figure 7.25, Daniels puts more shadows on Garbo's face by keying her from a cross-frontal position. Meanwhile, the low general illumination establishes the somber tone. In an impressive display of skill, Daniels darkens the tonality even more by periodically blocking and unblocking the keylight. This staccato flicker effect suggests the passing of the train, while making the tragic mood even more grim. The result is a very successful compromise, an expressive effect that allows us to see every detail of Garbo's performance.

Is this the lesson we should take from *Anna Karenina*—that all classical films are the same, while each unclassical film is unclassical in its own way? Certainly not. We must remember that the conventions of Hollywood lighting were often conventions of differentiation, encouraging cinematographers to treat different stories in different ways. Some films will call for a greater emphasis on genre; other films will call for a greater emphasis on glamour. *Mata Hari* (George Fitzmaurice, 1931) is about a famous spy; fittingly, Daniels relies more heavily on the genre/scene conventions of melodrama, using incredibly dark shadows for a scene when important secrets are stolen. By contrast, Ernst Lubitsch's *Ninotchka* (1939) features the bright tonalities of the comedy. Even here, Daniels still manages to work in an occasional effect, as when he uses a top-light to darken the mood for one of the film's rare dramatic scenes. For a classical

Figure 7.25 Greta Garbo's face goes from light to shadow and back again during the climactic suicide scene.

cinematographer like William Daniels, each film presents a unique problem, calling for a unique solution.

◎ ◎ ◎

The decade of the thirties is often seen as the high point of the classical
Hollywood style, and now we can say why: in the years following the transition to sound, Hollywood cinematographers perfected the art of balance.
In films like *Anna Karenina* (or *Peter Ibbetson*, or *Barbary Coast*, to take a
few more examples from 1935) the style modulates from one scene to
another, sometimes even from shot to shot, favoring expressivity at one
moment, and then glamour the next, yet always offering a thoughtful interpretation of the unfolding story. This is Hollywood lighting at its most
classical: never neutral, but always subtle.

In the following decade many cinematographers would continue to
practice the art of balance—in fact, many still do. However, while the classical style did not come to an end, it did continue to change, as cinematographers rebelled against certain principles or adapted others. As usual,
one impetus for change was technology. Three-color Technicolor came to
feature-film production in 1935, requiring cinematographers to rethink
the balance of narrative and spectacle while exploring the possibilities
and limitations of a new technology. Historically, this was followed in the
forties by the trend toward deep-focus cinematography, which responded
to new ways of thinking about realism. Technology aside, more and more
cinematographers began to question the ideal of balance, as more mannered approaches grew even more popular during the forties, culminating in the highly expressive style of the film noir. Part III will examine all
of these developments in more detail. Hollywood had crafted some remarkable patterns of shadow in the classical era, but those patterns would
continue to shift.

Part III
Shifting Patterns of Shadow

8 | The Promises and Problems of Technicolor

George Fitzmaurice's 1924 film *Cytherea* featured sequences photographed in Technicolor's new two-strip process. The film's cinematographer, Arthur Miller, had never photographed a color film before, and the Technicolor Corporation provided a cameraman, J. A. Ball, to collaborate with him on the color sequences. In an interview, Miller once described his experiences with Ball: "He said, 'Well, can you put a little more light under this couch.' It was right near the edge of the set, and I said, 'I can do it, and I will, but it usually isn't very bright under a couch. He said, 'I have to go on the theory that where there's no light, there's no color.' So he wasn't so dumb."[1] For Miller, lighting had to be guided by the norms of plausibility. For Ball, lighting had to be guided by a different objective: the need to showcase Technicolor's capabilities.

Historians agree that Technicolor photography posed a number of problems for the Hollywood style. In one early essay, Edward Buscombe argued that color was a threat to the (thoroughly conventional) realism of black-and-white films.[2] More recently, Scott Higgins has argued that the desire to put Technicolor on display could have conflicted with the need to use color in ways that were consistent with the classical norm of unobtrusive storytelling. Higgins's account is particularly useful because he explains the conflict in institutional terms, comparing the institutional agenda of the Technicolor Corporation with the ideals of classicism, as they had been articulated by institutions like the Society of Motion Picture Engineers and the American Society of Cinematographers.[3]

In chapter 5 I proposed that we think of the groups in Hollywood as a set of overlapping circles. The American Society of Cinematographers and the Society of Motion Picture Engineers shared several goals and ideals, but the overlap between them was never perfect, as each institution would prioritize those ideals in a manner consistent with its own institutional agenda. In this chapter I will argue that we can think of Hollywood's adoption of Technicolor in the same way. On the one hand, use of Technicolor assumed many of the ASC's cinematographic ideals, encouraging cinematographers to think of color as an extension of their existing practices. On the other hand, the Technicolor Corporation subtly modified those ideals to emphasize color's distinctive properties, thereby creating a new prioritization of functions. As in black-and-white cinematography, the result was a range of stylistic options, with some cinematographers attempting to create a new classical balance, and others opting for a more mannered approach by exploring color's potential to create unusual pictorial effects.

We can draw an analogy between the evolution of the discourse of Technicolor and that of the ASC. For instance Technicolor initially adopted an eclectic approach, willing to use any rhetorical strategy to advance the use of its namesake product. Later, as Technicolor moved closer to the center of Hollywood norms, its discourse became more classical, although the overlap would never become perfect.

As an example of the eclectic stage, we might consider a 1927 paper, presented to the SMPE by L. T. Troland. Here Troland offers a varied set of arguments in favor of Technicolor's two-strip process. First, Technicolor will help filmmakers tell certain kinds of stories: namely, stories are about gorgeous people in beautiful spaces. Second, he expands this argument by saying that Technicolor can enhance all kinds of stories, since all stories can benefit from an increased illusion of presence. He writes, "Now I believe that it is a psychological truth that the arousal of interest and emotion depends always upon some degree of *conviction*. A mere idea is not sufficient; it must be *'believed in.'* . . . Once we are embarked on this line of thought it is impossible to avoid accepting natural color."[4] These arguments suggest that Troland was already tailoring his rhetoric to suit the needs of a classical aesthetic. However, a few paragraphs later, Troland makes a completely different kind of argument. Discussing a hypothetical example involving the shift from black-and-white to color, he writes:

> Even under these circumstances it can be presumed safely that the distraction of attention results from the pleasantness or entertainment value of color *per se*, and hence we might argue that temporarily the dramatic action might be suspended without any net loss of interest. . . . If a picture is shown of a

woman in a gorgeous costume or with beautiful jewels, the audience should be given some time to appreciate the display before the story proceeds.[5]

Contradicting his own earlier arguments, Troland argues that narrative should be suspended from time to time, to allow audiences to appreciate some moments of Technicolor spectacle.

Seven years later, the Society of Motion Picture Engineers held a meeting to discuss Technicolor's new three-color process, which was to be introduced to feature filmmaking with *Becky Sharp* (Rouben Mamoulian, 1935). Natalie Kalmus, the ex-wife of Technicolor founder Herb Kalmus, was the company's leading color consultant, and she took the opportunity to deliver a now well-known address on the art of color, entitled "Color Consciousness." Like Troland, Kalmus offers an eclectic list of functional benefits, but a closer look reveals that Kalmus has modified the rhetoric of Technicolor, incorporating some familiar ideas from established Hollywood institutions like the SMPE and the ASC. In one passage, for instance, she invokes the SMPE's favored theme of technological progress as a step-by-step process, moving closer and closer to the goal of representing reality: "From a technical standpoint, motion pictures have been steadily tending toward more complete realism. . . . The advent of sound brought increased realism through the auditory sense. The last step—color, with the addition of the chromatic sensations, completed the process. Now motion pictures are able to duplicate faithfully all the auditory and visual sensations." In the very next paragraph, Kalmus shifts from the rhetoric of technology to the rhetoric of art: "A motion picture, however, will be merely an accurate record of certain events unless we guide this realism into the realms of art. To accomplish this it becomes necessary to augment the mechanical process with the inspirational work of the artist."[6] If the first passage was well suited to the pages of the *Journal of the Society of Motion Picture Engineers*, this second passage would not have looked out of place in *American Cinematographer*. Technology provides the appropriate mechanical tools, but a work of art must offer more than merely mechanical reproduction.

Kalmus then explains how Technicolor might enhance the artistry of a film. She points out that color can improve the sense of depth by separating the foreground from the background, and that color's visual variety can help a film maintain audience interest. Significantly, she spends several paragraphs explaining one particular function: the use of color to evoke a mood that is appropriate to the story. She writes, "Just as every scene has some definite dramatic mood—some definite emotional response which it seeks to arouse within the minds of the audience—so, too, has each scene, each type of action, its definitely indicated color which harmonized with that emotion."[7] There are clear affinities between this passage and the writings of Victor Milner, who had repeatedly invoked the right-mood-for-the-story theory in his various articles for *American Cinematographer*. For Kalmus, a skillful filmmaker will find correlations between graphic components and narrative events. She explicitly states that the filmmaker should be guided by the psychological associations of color, even as she notes that some colors produce a wide range of associations. Red, for instance, is associated with such disparate themes as danger, love, and revolution, while yellow evokes light, harvest, and jealousy. At the same time, Kalmus insists, "We must constantly practice color restraint."[8] If Technicolor is a tool for storytelling, then it does not need to be flamboyant to be artful. The artfulness of color is measured by its harmony with the mood of the story. Kalmus's tone echoes Milner's insistence that that mood lighting should work unobtrusively, at an unconscious level. The rhetoric of Technicolor has converged neatly with the existing Hollywood norms.

Of course, this should not suggest that Technicolor was perfectly integrated into the Hollywood system after 1935. Scott Higgins has explained how Technicolor experimented with a variety of different approaches during the thirties. First, Technicolor employed a "demonstration mode," using forceful color palettes in films like *La Cucaracha* (Lloyd Corrigan, 1934) and *Becky Sharp* to advertise what three-color Technicolor could do. When *Becky Sharp* turned out to be a disappointment, Technicolor adopted a "restrained mode," attempting to prove that color could be an unobtrusive addition to the Hollywood film. During this period, major studios began to use the technology in their films, such as *The Trail of the Lonesome Pine* (Henry Hathaway, 1936) and *A Star Is Born* (Wellman, 1937). By 1938, use of Technicolor was still rare, but it was becoming more widely accepted as a legitimate option for major studio films. Technicolor took the opportunity to introduce a more "assertive mode" of cinematography in *The Adventures of Robin Hood* (Michael Curtiz and William Keighley, 1938). Like the films of the demonstration mode, *Robin Hood* employs a color palette that is both varied and vivid. Like the films of the restrained mode, the film works hard to integrate the rainbow palette with the changing demands of the unfolding narrative.[9]

Higgins's subtle analysis reveals that the process of integration was complex, as filmmakers struggled to balance the demands of Technicolor with the demands of Hollywood. Significantly, these negotiations ended up proving a point that cinematographers had been trying to make for some time: that it is not always necessary to choose between spectacle and story. Sometimes it is possible to have both at the same time.

◎ ◎ ◎

During the second half of the thirties, David O. Selznick played a crucial role in endorsing the artistic possibilities of Technicolor by using the process in a number of major films, including *The Garden of Allah* (Richard Boleslawski, 1936), *A Star Is Born*, and, most famously, *Gone with the Wind* (Victor Fleming, 1939). Selznick was well known for writing memos to his employees, and these memos can greatly enrich our understanding of Hollywood's careful and occasionally contradictory adoption of Technicolor.

One very obvious source of conflict that Selznick needed to address was the struggle for authority over the camera crew. Just as cinematographers had tussled with engineers after the transition to sound, cinematographers found themselves struggling with Technicolor personnel during the transition to color. Originally, the Technicolor Corporation insisted that Hollywood producers use Technicolor cinematographers on their productions, since Technicolor cinematographers could provide a proficient level of technical skill, while foregrounding the color design in a way that was advantageous to the company. Meanwhile, many producers preferred working with the established masters of black-and-white cinematography, who had already demonstrated their ability to balance the demands of glamour, mood, and narrative clarity. As a result, most Technicolor films from the thirties were shot by two or more cinematographers—a situation that caused a number of on-set disputes about the best way to handle lighting and composition. In his early Technicolor production *The Garden of Allah*, Selznick endeavored to find the right combination of cinematographers for the job. He had originally wanted to use Charles Lang, a Paramount cinematographer expert in getting glamorous images of the film's star, Marlene Dietrich.[10] When he was unable to get Lang, Selznick borrowed MGM's Harold Rosson to work with Howard Greene, a Technicolor cinematographer. This arrangement caused considerable confusion on the set. In one of his memos, Selznick wrote:

> The Technicolor people told Hal Rosson that they hoped he understood that he was not the head cameraman on the picture, but that Green[e] was; and that

his, Rosson's, job was merely to watch the closeups of Miss Dietrich. This is, of course, not our intention and it promises trouble.

I wish you would immediately advise me whether or not there is anything in the Technicolor contract that prevents our putting on a director or supervisor of photography, such as Rosson, with authority over the Technicolor man. . . . I think on exteriors he [Greene] will probably do a finer job than any other black and white man could do. Furthermore, we certainly need him because while Rosson should be able to enormously improve the photography, without Green[e], Rosson would be lost.[11]

The Technicolor Corporation had more faith in its own cinematographer, Greene, though it acknowledged that he might need some help with the glamour shots. Selznick trusted Greene with the exteriors, but he still wanted veteran black-and-white cinematographer Hal Rosson to supervise all the photography. Both men were assigned to the film; not surprisingly, they soon began competing for authority. In a handwritten response to the memo quoted above, Boleslawski sided with Greene, while noting that Rosson had a "slight Napoleonic complex." Boleslawski proposed a solution: "I also suggested to Rosson to fall in love with Marlen[e]—in which case the billing and title would not matter because it all will be a work of heart, art and love—Rosson said he will seriously consider it."[12] Unfortunately, Boleslawski was unable to solve the problem. In a later memo Selznick criticized Rosson's work, while wondering if exterior expert Greene could be replaced by Technicolor cinematographer William Skall when the production switched to interiors.[13] In the end, Rosson and Greene shared a special Academy Award for their work on *The Garden of Allah*, but they never worked together again.

Selznick was famous for hiring and firing people to get the precise combination of effects that he wanted, but the conflicts that characterized Selznick's productions were not just the result of his demanding personality. They were the result of the system itself: a system that called on filmmakers to balance a wide range of competing functional demands. A few years later Selznick produced *Gone with the Wind*, an incredibly ambitious film that attempted to advertise the advantages of glorious Technicolor while keeping the audience engaged with a long and complex storyline. True to his thorny reputation, Selznick fought with the Technicolor consultants, and the nature of the struggle is very revealing. We might expect Technicolor to have advocated a bold color palette, to better demonstrate the beauty of their product. However, the situation was the reverse. Still committed to the "restrained mode," the Technicolor consultants encouraged Selznick to use more desaturated colors, and Selznick had to insist on a more eye-catching palette. One memo is worth quoting at length:

> We should have learned by now to take with a pound of salt much of what is said to us by the technicolor experts. I cannot conceive how we could have been talked into throwing away opportunities for magnificent color values in the face of our own rather full experience in technicolor, and in the face particularly of such experiences as the beautiful color values we got out of Dietrich's costumes in "The Garden of Allah," thanks to the insistence of Dietrich and Dryden, and despite the squawks and prophecies of doom from the technicolor experts.
>
> Further, the color values of "The Garden of Allah" had no comparable dramatic significance, whereas the proper telling of our story involves a dramatic and changing use of color as the period and the fortunes of the people change.
>
> Examine the history of color pictures: The one thing that is still talked about in "Becky Sharp" is the red capes of the soldiers as they went off to Waterloo. What made "Cucaracha" a success, and did so much for the technicolor company, were the colors as used by Jones for his costumes. The redeeming feature of "Vogues" was the marvelous use of color in the women's styles. The best thing about the "Follies" was the beautiful way in which colors of sets and costumes were blended, as in the ballet.
>
> I am the last one that wants in any scene a glaring and unattractive riot of color—and I think I was the first to insist upon neutralizing of various color elements, particularly of sets, so that the technicolor process would not obtrude on dramatic scenes, but I certainly never thought that this would reach the point where a sharp use of color for dramatic purposes would be completely eliminated; nor did I ever feel that we were going to throw away the opportunity to get true beauty in a combination of sets and costumes.[14]

Selznick argues that it would be a waste of resources to use Technicolor without fully exploiting it for its spectacular qualities. However, he is not calling on the filmmakers to emphasize color simply for its own sake. He reasons that there are a number of scenes where bold colors are justified on dramatic grounds. Specifically, he proposes that the film should start out with saturated colors, and then switch to a more desaturated scheme when the mood grows more somber. The result will be a truly multifunctional use of color: dramatic and spectacular at the same time.

While this memo is primarily concerned with production design, Selznick was also excited about color's potential contribution to the art of lighting. Unlike black-and-white, lighting in Technicolor could use different hues to indicate various time-appropriate effects, such as the effect of candlelight. Although most Technicolor cinematographers preferred working with bright levels of illumination, Selznick actively encouraged his cinematographers to work with shadows, to more accurately indicate what it was like to live in a world illuminated by candlelight.[15] Unfortunately, most of his cinematographers took the recommendation too far, creating dark pools of shadow that threatened to swallow up the performers. Selznick complained:

> It may be my fault for insisting on effect photography but I cannot seem ever to be able to drive into people's heads that when I ask for effect photography I do not mean that the whole screen should be so dark that we cannot tell what is going on. . . . I would appreciate it if you would get together with Garmes and Renihan [sic] and make clear that we simply cannot tolerate any more photography that is so dark as to bewilder the audience—and that if they cannot get effect lighting without avoiding this risk they should forget the effect lighting and light everything as though it were a newsreel. If we can't get artistry and clarity, let's forget the artistry.[16]

Soon after this Selznick replaced Lee Garmes with Ernest Haller. Fortunately Haller and Rennahan did not shoot *Gone with the Wind* like a newsreel. The film is filled with superb examples of effect-lighting. Indeed, following Higgins, we can see *Gone with the Wind* as the culmination of Hollywood's efforts to integrate Technicolor into the Hollywood style. Selznick may have implied that clarity was more important than artistry, but it is clear that he really wanted both. Technicolor lighting was also an art of balance, aiming to achieve multiple functions at the same time.

Turning from the case of Selznick to the development of Technicolor lighting as a whole, we note that Selznick's efforts took place within the context of the increasing conventionalization of Technicolor lighting. As was the case with black-and-white lighting in the silent period, certain techniques were well on their way to becoming default conventions, favored by a wide range of cinematographers because they could accomplish a variety of functions in an efficient manner.

As early as 1935, C. W. Handley had suggested that cinematographers use tungsten incandescent units for effect-lighting.[17] Technicolor film stock was balanced for daylight, so incandescent light registered as orange. Cinematographers relied on daylight-balanced arc lights as the default norm for Technicolor, but cinematographers quickly realized that they could take advantage of this difference in color temperature to improve certain lighting effects. In interiors, an incandescent might be used to add an orange glow to candlelight, gaslight, and fireplace effects. In exteriors, an incandescent light could suggest the warm rays of a sunset.

When combined with a blue gel, an arc light could create a very compelling beam of moonlight streaming through a window. One popular technique was to combine blue and orange lights within a single shot, suggesting a candlelit room with moonlight streaming through the window. Meanwhile, cinematographers learned how to take advantage of the arc light's hardness, which made it a great tool for producing cast shadows. As an example of their experimentation, cinematographers even began to create cast shadows in unusual ways, using "doubles" to produce the shadows. For instance, in *For Whom the Bell Tolls* (Sam Wood, 1943), Ray Rennahan photographs Gary Cooper with a cast shadow in the background—but closer observation reveals that the shadow has been created by someone else, copying Cooper's gestures to create the illusion that the shadow belongs to Cooper. Although this is a somewhat unusual example, it suggests the degree to which cinematographers were willing to use crafty lighting effects for pictorial purposes.

Perhaps the most common lighting effect in Technicolor was the silhouette. Ever since filmmakers like Maurice Tourneur had popularized the technique in the silent period, the silhouette was seen as a simple way to add a dash of spectacle to any film. Color heightened the spectacle even more, and the image of a silhouette situated against a vividly colored sky became a Technicolor cliché, appearing in a wide range of films, such as *Gone with the Wind* and *She Wore a Yellow Ribbon* (John Ford, 1949). While all of these conventions are merely variations on established black-and-white norms, they interpret those norms in a particular way, placing a special emphasis on spectacle, foregrounding the artistry of Technicolor without necessarily sacrificing the story.

At first glance, we might think that it was easy to adopt Hollywood's genre/scene conventions in the Technicolor context, since there are clear similarities between Natalie Kalmus's theory of "color consciousness" and the ASC's concept of the right-mood-for-the-story. However, there were a few complications involved in the transition. First, filmmakers had to determine if color was a good option only for certain genres, or in fact for all genres. In other words, some filmmakers believed that dramas and melodramas were more effective in black-and-white, while noting that color could be a good resource for the musical or the period picture.

Meanwhile, other filmmakers believed that all films would eventually be shot in color. They saw no reason why Technicolor could not be adapted to the somber style of the drama, or the stark style of the melodrama.

Here we should remember that the genre conventions could be applied at two levels: that of the film, and that of the scene. On the level of the film, it is certainly true that Technicolor was often confined to specific genres. Of the twelve films that won the Academy Award for cinematography during the period 1939–1950, eleven could be described as period pieces. (The only exception is *Leave Her to Heaven*, John Stahl's color-noir from 1945.) However, when we look at the level of the scene, we see that filmmakers were trying to make use of Technicolor in the full range of established genre/scene conventions. For instance, in *Down Argentine Way* (Irving Cummings, 1940), Ray Rennahan and Leon Shamroy employ gentle leaf patterns during a romantic scene. In the bullfighting drama *Blood and Sand* (Rouben Mamoulian, 1941), Rennahan and Ernest Palmer follow the melodramatic convention of casting shadows on a wall during a fight scene. We can even find mood lighting in the middle of a sunny MGM musical. An example from the end of the period under study in this book, *Summer Stock* (Charles Walters, 1950), features a wonderful image of a despairing Joe (Gene Kelly): the dim foreground looks even dimmer when compared to the magenta background (fig. 8.1, *color section*). A few moments later Joe steps out of the shadows and begins a dance routine; his move from the shadowed area in the foreground to the more colorful area in the background embodies the scene's movement from sadness to joy.

It should not surprise us to learn that cinematographers were eager to apply the existing black-and-white conventions to the color context. After all, the discourse of the ASC placed a great deal of emphasis on the expressive power of the genre/scene conventions. However, we have seen that Natalie Kalmus encouraged cinematographers to do something more: to use color itself as an expressive technique. To what extent did filmmakers seek to codify the emotional effects of certain hues? Did they always use red for scenes of anger? Did a yellowish green become the color of jealousy? The answer to all of these questions is probably "no." As Higgins has pointed out, Mamoulian experimented with obvious color patterns in *Becky Sharp*, but the film's style was too self-consciously theatrical to work within the existing Hollywood norms. Most future Technicolor filmmakers would use color associations with more subtlety.

Figure 8.1
Gene Kelly's face is in shadow in a somber moment from *Summer Stock* (1950).

Figure 8.2
A variety of saturated hues for a modeling scene in *Cover Girl* (1944).

Figure 8.3
Cover Girl: Muted hues in a more serious scene.

Figure 8.4
Bright illumination allows the colorful costumes to stand out in *That Night in Rio* (1941).

Figure 8.5
The composition emphasizes Marlene Dietrich's eyes, lips, and hair in *The Garden of Allah* (1936).

Figure 8.6
A kicker emphasizes the grotesque expression of a man in the same scene.

Figure 8.7
King Solomon's Mines (1950): Stewart Granger is shot on location.

Figure 8.8
The corresponding shot of Deborah Kerr was shot in a studio.

Figure 8.9
A semi-silhouette from *The Garden of Allah*.

Figure 8.10
A complicated lighting scheme with several moving lights, from *The Black Swan* (1942).

Figure 8.11
The Black Swan: An elaborate composition with a variety of cast shadows.

Figure 8.12
Two different color temperatures on Gene Tierney's face in *Leave Her to Heaven* (1945).

Nevertheless, we should note that filmmakers did not abandon the idea that graphic qualities could produce fairly predictable emotional associations. While they may have rejected associations at the level of hue, they probably accepted associations on other levels. A happy scene might have a variety of hues, at a high level of saturation, whereas a somber scene might employ a single desaturated hue. Charles Vidor's *Cover Girl* (1944) implicitly comments on this strategy: for the happy scene when Rusty models for the first time, a montage shows her being illuminated with a variety of colorful lights. In one shot, the photographer and the publisher stand in front of a blue background, near a yellow lamp in the middleground, as a man walks by in the foreground carrying a green and magenta filter. This image is intercut with a shot of Rita Hayworth wearing a light brown dress, with a green lamp appearing against a green wall in the background, as a man carries a red box in the foreground (see fig. 8.2, *color section*). A few minutes later, the mood becomes more serious, and a dramatic scene is played entirely in dull shades of brown. In good times Rusty is surrounded by a cheery cacophony of saturated colors; when she is feeling down, she is surrounded by drab earthtones (see fig. 8.3, *color section*). In these two scenes cinematographers Allen Davey and Rudolph Maté are simply applying the same principles that David O. Selznick had advocated during the filming of *Gone with the Wind*. Before the war, Tara looks bright and vivid; later, when Scarlett returns to Tara and finds that her father has gone insane, the colors have muted, becoming dull and gray.

The control of saturation and hue was also an important factor in composition. Here the challenge was to balance a variety of functions: creating the illusion of depth, keeping the audience's attention on the story, and taking advantage of Technicolor's ability to produce beautiful images. Most cinematographers believed that Technicolor was very well suited to creating the illusion of depth. As Arthur Miller once said, "What makes it easy in color is if you put green against blue, you've got separation. In black and white, you have to create these separations with your lighting; that's what separates the men from the boys."[18] In other words, in Technicolor a cinematographer did not have to worry about backlighting every detail, as he would with black-and-white. Instead, he could rely on differences in hue to separate the planes. To facilitate this process, Technicolor consultants like Natalie Kalmus often recommended using neutral backgrounds, allowing the more saturated costumes of the performers to pop. Still, in order to direct attention, cinematographers typically returned to the black-and-white convention of lighting the foreground a bit more brightly than the background during a dramatic scene. For a more spectacular scene, such as a musical number, the cinematographer might flood the set with light, dispersing the audience's attention across the frame. In figure 8.4 (*color section*), from *That Night in Rio* (Irving Cummings, 1941), Leon Shamroy's bright illumination allows the colorful costumes to dazzle the eye.

In exterior shots Technicolor posed a new problem for composition: How should the cinematographer photograph the color of the sky? The temptation was to enhance the pictorial beauty of the shot by photographing the blue skies at a high level of saturation. Unfortunately, this choice ran the risk of upstaging the actors, as even the most gregarious scene-stealing ham could not compete with the vivid blue background. Some filmmakers chose to err on the side of spectacle, as in *For Whom the Bells Tolls*, where the distracting blue skies serve as an overly pretty background for a somber war story. Other filmmakers preferred a more restrained approach, using filters and framing to produce desaturated skies, thereby allowing the actors to command the greatest screen presence.

Of all the conventions, Technicolor may have had the most impact on those of figure-lighting. To be sure, some historical norms were preserved. For instance, most cinematographers continued to use frontal keys for women, while employing cross-frontal keys for men. This technique would model the features of the man, while smoothing the features of the woman. However, technical constraints made it difficult to adopt the preexisting conventions without a revision of technique. Technicolor film stock was very slow, requiring high levels of illumination. As J. A. Ball pointed out, some cinematographers would make the mistake of flooding the set with light, thereby destroying the modeling.[19] In addition, cinematographers had to take color temperature into consideration. For reasons of efficiency, Technicolor cinematographers relied heavily on arc lights, but these units were harder than incandescent lights, which were the more common choice for black-and-white filming. The hardness of an arc light could be an advantage when photographing a rugged male star, but it was a disadvantage when photographing women. A cinematographer might need to compensate by using more diffusion, either on the lamps or on the lens.

More important, Technicolor required cinematographers to rethink a topic that they often took for granted: skin tone. In black-and-white films, cinematographers usually took white skin as the norm, employing lighting and laboratory practices that could reliably produce a certain image of whiteness. As Richard Dyer has persuasively argued, "In the history of photography and film, getting the right image meant getting the one which conformed to prevalent ideas of humanity. This included ideas of whiteness, of what colour—what range of hue—white people wanted white people to be."[20] The resulting conventions were often split along the lines of sexual difference: men could look somewhat tan, but women usually had glowing alabaster skin. This contrast was particularly strong during the silent period, when stars like Lillian Gish embodied a purified ideal of womanhood. As Dyer notes, "The angelically glowing white woman is an extreme representation, precisely because it is an idealization."[21] By the thirties, this ethereal ideal had come to appear overly sentimental. Cinematographers began using fewer layers of gauze for their female close-ups, but they continued to follow the established skin tone conventions, maintaining the white ideal as the norm by creating moderately light skin tones for most men and virtually transparent skin tones for most women.

These conventions simply did not work in Technicolor. J. A. Ball explains the situation that cinematographers were facing:

> In color photography, all very full exposures tend to bleach out to white, and all low exposures tend to drop into black. A highlight upon a face in black-and-white photography can, in the final print, be merely the bare celluloid, and the result will be still entirely satisfactory; but if, in a color print, such a condition exist, the delicate flesh tint will, in that area, be bleached out to white, and the face will look blotchy. All areas of the face should, therefore, be reproduced in such a manner as to yield a good flesh tint.[22]

On one level this is a technical problem. Because Technicolor did not have the latitude of black-and-white stock, cinematographers had to be particularly careful with their exposures. However, Ball's solution to this technical problem involves a shift in representational ideals, to one favoring healthy-looking saturated skin tones over purified white skin tones.

Whereas the ASC usually left its assumptions about skin tone unspoken, the Technicolor Corporation actively promoted its new approach to skin tones as a selling point. Even during the two-strip period, L. T. Troland had argued, "Undoubtedly the greatest 'kick' of color, at least for the male members of an audience, consists in the value which it adds to the delineation of feminine beauty. All pretty girls in black and white are pale and consumptive."[23] Technicolor promised to deliver a new perspective on female beauty, offering the image as a spectacle to be appreciated for its own sake. Later, three-strip Technicolor films would flaunt the process's ability to capture a wide range of skin tones. From *La Cucaracha* (Corrigan, 1934) to *King Solomon's Mines* (Compton Bennett and Andrew Marton, 1950), filmmakers relied on "exotic" subject matter to put the technology's skills on display, turning racial diversity into pictorial spectacle.

Typically this exoticism would function to define whiteness as the norm. In figure 8.5 (*color section*), from *The Garden of Allah*, the top-frontal key-light brightens and smooths the facial features of Marlene Dietrich, allowing her red lips and blue eyes to stand out. Her dark blue outfit emphasizes the lightness of her face, while the yellowish backlight brings out the blonde in her hair. These strategies already work to emphasize the whiteness of her character, but that racial identity is underlined every more strongly when the film cuts to the shot in figure 8.6 (*color section*). With a bright kicker and a prominent eye-light, the lighting works to emphasize the man's grotesque expression, thereby marking him as the pictorial opposite of Dietrich's beauty. Two boys stand behind him, for no other purpose than to allow the filmmakers to underscore Technicolor's ability to capture a wide range of skin tones. The result is a kind of racial montage, domesticating the exotic presence of Marlene Dietrich through a strategic use of contrast.

More than a decade later, MGM sent a crew to Africa to film *King Solomon's Mines* (1950). Again, the filmmakers have worked to emphasize the whiteness of the female lead (Deborah Kerr) through contrast—contrast with the tan skin of the male lead (Stewart Granger), and contrast with the various African extras shown throughout the film. The filmmakers were even willing to put continuity at risk to ensure that Kerr would be seen with maximum glamour. In one scene the establishing shot shows Kerr and Granger together on location in Africa. Granger's single, shot on location, is then intercut with Kerr's single, shot in a studio. Figure 8.7 (*color section*) shows Granger's single: a simple set-up using the sun as a high cross-frontal key. The contrast with Kerr's single is striking: three-point lighting provides gentle modeling on Kerr's face, which looks particularly delicate against the vivid studio sky (fig. 8.8, *color section*).

Technicolor had to balance its interest in racial diversity as a form of pictorial spectacle with its commitment to narrative clarity—a commitment that required a certain amount of racial homogeneity. Winton Hoch, a Technicolor cinematographer, explains:

> If one will note the varying complexions of people, he will readily appreciate that if three or four persons were lined up side by side to be photographed, it

214

would be highly desirable and probably very necessary to correct the flesh tones and greatly reduce the tone spread. This must not be interpreted as meaning that all flesh tones should appear alike. Variations of tone are very desirable. It is the extremes that are undesirable. Obviously a white man with a heavy tan who photographs like an Indian is not a very convincing white man.[24]

On the one hand, Hoch wanted to capture certain variations in skin tones, thereby demonstrating Technicolor's ability to render details that a black-and-white cinematographer could not even consider capturing. On the other hand, certain narratives demanded that the race of each character be clearly marked. Hoch suggested that each race be represented with a stable set of nonoverlapping tonalities in order to do so.

◉ ◉ ◉

In short, the technical problems of Technicolor were also ideological problems. It was not simply a matter of translating the black-and-white figure-lighting conventions into the terms of color cinematography. Rather, the technical resources of color had to be used to create patterns of similarity and difference—patterns that were often defined by social categories like race and gender.

Looking at the above examples of lighting conventions, it is tempting to conclude that Technicolor had been *contained* by a process of conventionalization. The new technology represented a threat to the Hollywood norms; Hollywood responded to that threat by assimilating it. While there is some truth to this account, it must be qualified in an important way. The Hollywood norms were never stable. They were created by a variety of competing institutions, and each institution interpreted those norms in slightly different ways. Because the Technicolor Corporation wanted to do business with Hollywood, it necessarily adopted some of Hollywood's ideals. But the addition of a new institution to our pattern of overlapping circles inevitably shifted the center of the pattern. Technicolor gave cinematographers a new way to think about the art of lighting. They were still committed to storytelling, but they could support the story in a whole new way: calling attention to an essential plot point with a burst of red, or establishing atmosphere with a flickering orange fireplace, or expressing joy with a splashy display of hues. Technicolor provided a new set of tools that stimulated creativity, reminding cinematographers that storytelling could be spectacular.

Still, most Technicolor cinematographers were classicists, finding compromise solutions that allowed them, as ever, to balance a wide range of potentially competing goals. For instance, most cinematographers would never shine a garishly colored light on an actor's face. To do so would violate the demand for appropriate skin tones. At the same time, cinematographers did not want to abandon the option of using colored gels, a valuable item in the creative toolbox. A popular compromise solution was to use balanced light on the actor's face while using a colored gel on the backlight. In *For Whom the Bell Tolls*, one scene shows Gary Cooper and Ingrid Bergman facing each other while standing next to a fireplace. Rennahan uses a golden flickering backlight on both characters. This solution is not entirely logical, since there is only one fireplace and it could not possibly light both characters from behind. However, this small sacrifice in continuity allows Rennahan to preserve the skin tones of both performers while adding the artistic detail of the flickering fireplace effect. Similarly, both *Cover Girl* and *Summer Stock* feature scenes that integrate gelled lamps into the story: in the former, the protagonist goes to a modeling studio with photographic lighting equipment; in the latter, Gene Kelly's character (Joe) moves around some lighting equipment while singing a romantic song. In both cases the cinematographers could have justified some very bold effects, but they settled for the compromise solution, using gelled lights as backlights while employing balanced lights for their keys.

We have also seen that filmmakers routinely took advantage of Technicolor's affinity for rendering skies by shooting spectacular silhouette shots. However, this does not mean that filmmakers abandoned storytelling for spectacle. Rather, they combined the two functions by making subtle adjustments to their lighting schemes when necessary. *The Garden of Allah* features a spectacular scene showing Marlene Dietrich and Alan Marshal silhouetted against the desert sky. When the film cuts in for a closer view, Rosson and Greene cheat by illuminating Marshal's face with a cross-backlight, along with some subtle fill from the side (fig. 8.9, *color section*). The result is a small sacrifice in continuity in return for a significant gain in narrative clarity—and no loss in pictorial beauty.

Cinematographers could also combine storytelling and pictorial beauty by saving their most striking shots for key dramatic moments. Indeed, Technicolor was a superb medium for the strategy of selective emphasis. For instance, in *The Yearling* (Clarence Brown, 1946) Charles Rosher uses a remarkably restrained color palette throughout, favoring a tasteful mixture of browns and greens to portray the family's austere forest home. This visual restraint makes it all the more effective when he reveals some of his bluest skies during the joyous scene when the boy first gets the faun. Going in the opposite emotional direction, in *Captains of the Clouds* (Michael Curtiz, 1942), Sol Polito relies on a colorful high-key for the bulk of the film, saving his darkest shadows for some somber moments during the final battle. By manipulating the color palette, a cinematographer could build a significant macro-structure based on the decisions made at the micro-level of the scene or even of the individual shot.

As in black-and-white cinematography, the selective emphasis strategy carried certain risks: used poorly, it might appear artificial and distracting. The strategy is particularly obvious in films that combine Technicolor with black-and-white, as in *The Little Colonel* (David Butler, 1935), which employs a short Technicolor sequence at the end to emphasize the happiness of the conclusion. Even films that are shot entirely in Technicolor can provide some rather aggressive transitions. *Becky Sharp* provides a well-known example. After a series of scenes with relatively restrained color, the film shifts to a much bolder palette during the Waterloo sequence. Characters rush across the screen in primary colors, such as blue and red, while a low-placed light casts their shadows on a background wall. The tension between dramatic emphasis and pictorial display is very strong in this scene. The director, Rouben Mamoulian, claimed that dramatic emphasis was his first priority. He even boasted, "You will see how inconspicuously, but with telling effect, the sequence builds to a climax through a series of intercut shots which progress from the coolness and sobriety of colors like gray, blue, green, and pale yellow, to the exciting danger and threat of deep orange and flaming red."[25] However, the scene leaves no lasting impression as the film's narrative climax; rather, it is best remembered as a circus of visual spectacle, putting Technicolor's palette on vibrant display.

Most filmmakers preferred subtler solutions, using a gradual shift to ease the transition. In *Gone with the Wind* the filmmakers opt to use heavy shadows for the scene in which Scarlett and Rhett take Melanie and her newborn baby to the carriage. Rather than cut directly to an image with the darkest possible tonalities, the filmmakers steadily shift the tonalities, moving subtly from semidark shadows to nearly impenetrable darkness. The scene has all the expressivity of the *Becky Sharp* scene, but the gradual progression keeps our attention on the highly dramatic events in the unfolding narrative, rather than luring our attention with a parade of garish costumes.

In addition to manipulating the lighting, a cinematographer could create a gradual shift in tone by asking the laboratory to brighten some scenes and darken others, thereby creating more nuanced transitions between shots. Taking both black-and-white and color cinematography into account, Technicolor founder Herbert Kalmus had this to say on the subject:

> There is the same problem in black and white photography, namely, if scenes very close to the light end of the printing scale and scenes very close to the dark end of the printing scale come eventually to be cut together in the same reel, it may well be that one or the other of them will have to compromise in the final printing with consequent possibility of the loss of a very fine effect. The remedy for this is to avoid a too extreme level of lighting in either direction and to allow for a little adjustment to be effected in the printing in the creation of effects or moods.[26]

These challenges were particularly severe in the case of three-color Technicolor stock, which had less latitude than black-and-white stock. However, the solution was essentially the same in either case. A skillful cinematographer would provide the laboratory with a thick negative, giving the laboratory more options in printing, allowing for some remarkably subtle transitions in tone.

In short, Technicolor may have shifted the balance of Hollywood ideals by moving pictorial spectacle closer to the center, but it did not prevent filmmakers from finding classical solutions to the problem of conflicting conventions. Films like *Gone with the Wind* and *The Yearling* are masterpieces in the art of balance, using attractive figure-lighting, nuanced lighting effects, and elegant compositions to tell their stories with a mixture of clarity, mood, and pictorial beauty. Still, it would be a mistake to conclude that all Technicolor films look the same, perfectly integrated with the Hollywood system. Many cinematographers were fascinated by the new possibilities of Technicolor, and they took the opportunity to experiment, creating a bold new style of Technicolor pictorialism. Leon Shamroy is a case in point. Shamroy was certainly capable of crafting works with a classical sense of balance; for instance, his somber photography for *Wilson* (Henry King, 1944) is perfectly suited to the film's serious dramatic subject. However, in films like *The Black Swan* (Henry King, 1942) and *Leave Her to Heaven* (John M. Stahl, 1945), Shamroy takes Technicolor lighting in a mannerist direction, pushing the pictorialism to the point where it dominates all the other functions.

Shamroy won his first Academy Award for *The Black Swan*, a pirate story with Tyrone Power and Maureen O'Hara (who would eventually come to be known as the Queen of Technicolor). In figure 8.10 (*color section*) Jamie Waring (Power) shares a room with Lady Denby (O'Hara) aboard a pirate ship. Shamroy takes advantage of the setting to create some remarkably complex effect-lighting. A blue arc light creates a window pattern behind O'Hara, and the window pattern jostles up and down to suggest the movement of the seafaring ship. Strictly speaking, this window pattern does not make a lot of sense, since we see the moon clearly located behind Power's head. This moon effect motivates a second blue window pattern, moving up and down on the right. Also on the right side, Shamroy has added a yellow light, to mimic the effect of the candle. This candle motivates the warm skin tones of Power and O'Hara. The shot is already pretty complex, but Shamroy adds one more touch: the shadow areas are illuminated by a light that appears to be reflecting off water, suggesting the presence of the sea. When we consider these effects individually, they are thoroughly conventional. Moonlight was typically blue, candlelight was typically orange or yellow, window patterns were typically created by arc lights, ship scenes typically had lights moving up and down, and ocean scenes typically had water effects. What makes the scene unusual is the fact that Shamroy has arranged all of these effects in concert. The result is an extraordinarily dynamic shot, with interesting visual details scattered all over the frame.

Similarly, in figure 8.11 (*color section*) Shamroy takes the idea of the daytime window pattern and pushes it to an extreme. By the time of this film, the window pattern was thoroughly conventional, a standard method of indicating the time of day. Here Shamroy uses the characteristic hardness of the arc light to creating a stunning series of patterns, complementing the straight lines of the room with the curved lines of the windows, while contrasting the tiny details on the right with the large flourishes on the left. As we have seen, Technicolor encouraged filmmakers to paint their backgrounds in neutral tones, allowing the actors to stand out. Shamroy has followed the letter of this rule, but he has violated the spirit: even the colorful costumes cannot prevent the actors from getting lost in this maze of shadows. Rather than creating a balance of functions, Shamroy has prioritized pictorialism above all else.

After using an expressive, somber style for *Wilson*, Shamroy returned to the mannered approach for *Leave Her to Heaven*. With a femme fatale protagonist and a plot featuring bizarre twists of fate, the film is sometimes described as a film noir in color. I think it is equally useful to see the film as an extension of the pictorial approach Shamroy had developed in *The Black Swan*—a decidedly un-noir film, despite the presence of the word "black" in the title. For instance, the trial scene reuses some of the set design details from figure 8.11, but Shamroy creates variety by using a new kind of window pattern: a trapezoidal pattern cast from above. In the New Mexico sequence, Shamroy experiments with orange gels, creating vividly colored light without necessarily motivating it as the light of a fire or candle. Most Technicolor cinematographers agreed that getting appropriate skin tones was their number-one priority. This is why the figure-lighting in figure 8.12 is so remarkable. Not content with restricting the application of an orange gel to the backlight, Shamroy emblazons the left side of Gene Tierney's face with an orange cross-light, flagging off the top to keep her forehead in shadow. Then he uses a strong top-frontal fill light, smoothing the features of her face while clearly defining her jaw-line and cheekbones. The image is certainly glamorous, but it is very unusual, placing two different color temperatures on Tierney's face. Shamroy could have made the image more conventional by insisting on a clear motivation for the orange light. While the film does establish the presence of a light source in the other room, the motivation is ultimately very weak for such an unusual effect. As in *The Black Swan*, the primary motivation seems to be pictorial, with Shamroy flaunting his ability to create new combinations of color.

Shamroy's own account is a good example of the mannerist sensibility, which rejects some Hollywood ideas while embracing others:

> I remember when I was making *The Black Swan*, I wanted to dispense with the usual Technicolor man from Kalmus, and to emulate the old masters, men like Van Dyke and Rembrandt, and I'd say to Zanuck, "When you're shooting a sunset, use yellow light instead of white light, and ignore realism, make a deliberate mistake." And Mrs. Kalmus went to Zanuck and said, "That isn't the way a color picture should be photographed." Zanuck stepped on me, but I still was the first black and white man to win an Oscar for color—with that picture.[27]

Shamroy's rhetoric is confident and rebellious, but we should also remember that he was a consummate company man who worked at Fox until the sixties. He is not a maverick trying to break every rule in the system. He breaks rules strategically, following some suggestions but not others. Shamroy hopes to break free from Kalmus's insistence on realism and restraint, but he does not call for a radical modernist aesthetic.

Instead, he hopes to paint with light, proposing Rembrandt as a model in the same way that Cecil B. DeMille had proposed him almost thirty years before.

⊚ ⊚ ⊚

We might explain these mannered touches by pointing to Shamroy's competitive personality. Although he won four Academy Awards, he claimed jokingly that he deserved an award for every single film he shot.[28] However, such a biographical explanation would miss an important point: competition was built into the system. Technicolor had adopted many of the rhetorical strategies of the ASC, and most cinematographers began to think of Technicolor as an opportunity to exercise, even flaunt, their artistic talents. Some cinematographers did so by practicing the art of balance, carefully crafting a multifunctional style. Other cinematographers, like Shamroy, proved their artistic worth by pushing the limits of one particular aesthetic function.

Hal Mohr once said that he did not distinguish between color cinematography and black-and-white cinematography. We should not take this to mean that Technicolor was simply assimilated unthinkingly into the pre-color Hollywood norms. The Technicolor Corporation worked hard to make that assimilation happen, and to make sure it happened on their terms. The end result may have been a convergence of styles, but that convergence was the result of years of strategy and struggle.

The Flow of the River

I n a beautiful passage from "The Evolution of the Language of Cinema," André Bazin writes:

> By 1939 the cinema had arrived at what geographers call the equilibrium-profile of a river. By this is meant that ideal mathematical curve which results from the requisite amount of erosion. Having reached this equilibrium-profile, the river flows effortlessly from its source to its mouth without further deepening of its bed. But if any geological movement occurs which raises the erosion level and modifies the height of the source, the water sets to work again, seeps into the surrounding land, goes deeper, burrowing and digging. Sometimes when it is a chalk bed, a new pattern is dug across the plain, almost invisible but found to be complex and winding, if one follows the flow of the water.[1]

Bazin's metaphor suggests a complex combination of continuity and change. Just as Bazin's river is recognizably the same river no matter what changes occur, the system of the classical Hollywood cinema continued to function smoothly throughout the forties. Yet when we take a closer look at the films, we find that the system has changed in many ways. For Bazin, the crucial change was the development of deep-focus cinematography. Other historians would point to equally important shifts: the gradual elimination of lens diffusion over the course of the decade; the turn to location shooting as an alternative to the studio; and, most famously, the heightened expressivity of the film noir style.

In this chapter I will examine these shifts, attempting to do justice to the nuances of Bazin's metaphor. The classical cinema had developed an elaborate set of conventions and functions. Over the years the conventions would adjust and transform, but cinematographers would continue to face the problem of conflicts, as they struggled to balance the core classical functions: storytelling, realism, pictorial quality, and glamour. The equilibrium-profile was shifting.

Many of the changes that occurred in cinematography during this period were similar to those that had taken place in the field of still photography over the prior few decades. There is no doubt that these developments in the history of photography had an influence on Hollywood lighting during this period, as cinematographers traded softness for sharpness. The challenge for us is to untangle the lines of influence. Perhaps one problem is that "influence" is a misleading term, one which proposes photography as cause and cinematography as effect. As the art historian Michael Baxandall has suggested, it can be useful to reverse the terms.[2] Rather than see cinematographers as passive adapters, we can think of them as active innovators, strategically drawing on a variety of photographic trends and transforming them to suit their own purposes. The results of their endeavors were complex: some cinematographers borrowed the new techniques to support classical storytelling ideals; other cinematographers tried to push classical filmmaking in a more mannered direction.

Looking at the field of photography, we can consider three distinct but related trends: straight photography, the New Deal–era photographs of the Farm Security Administration, and the photos found in new picture magazines like *Life*. Straight photography began in the teens, as a reaction against the excesses of pictorialism. Whereas pictorialist photographers used various softening techniques to make their work look more like painting, straight photographers firmly rejected the analogy with painting. According to California-based photographer Edward Weston, "The painter can eliminate or combine to suit his fancy. The photographer can, too, by messing around with a lot of Gum Arabic and paint and brushes, while destroying at the same time photography's chiefest charm, its subtle rendering of textures and the elusive qualities in shifting lights and shadows."[3] Following the principle of medium-specificity, Weston argues that the photographer should take advantage of the traits that make photography distinct from the art of painting, such as its delicate handling of textures, its astonishingly nuanced range of tonalities, and its ability to capture ephemeral moments.

Significantly, Weston does not reject the goal of promoting photography as an art. Indeed, his defense of photography is quite ambitious—he argues that photography can augment human vision, allowing us to see the details of the world in a way that no other medium can.[4] To accomplish this goal, Weston recommends the practice of deep-focus photography, as an explicit contrast to the soft-focus creations of the pictorialists: "Look then with a discriminating eye at the photograph exposed to view on the Museum wall. It should be sharply focused, clearly defined from edge to edge, from nearest object to most distant."[5] Deep focus was such an important strategy that Weston became a member of an informal group known as the "Group f/64," the term referring to an extremely narrow aperture opening (f-stop) that photographers used to create deep-focus images.

Paul Strand's defense of straight photography was even more ambitious. In a 1922 essay, "Photography and the New God," Strand argued that the recent trend toward straight photography was nothing less than the expression of a new attitude regarding the relationship between art and machines. According to Strand, modern life was in danger of being overwhelmed by a new cult of the machine—a cult that was threatening to make the artist irrelevant. At first glance, he argued, it would appear that photography was a symptom of the overmechanized culture of modernity. However, Strand goes on to suggest that this symptom might actually turn out to be a cure:

> Signs of this imperative revaluation of the idea of the machine are beginning to manifest themselves. And significantly, one might almost say ironically enough, not among the least important is the emerging demonstration on the part of the artist of the immense possibilities in the creative control of one form of the machine, the camera. For he it is who, despite his social maladjustment, has taken to himself with love a dead thing unwittingly contributed by the scientist, and through its conscious use, is revealing a new and living act of vision.[6]

The photographer proves that art is still alive, even in the midst of modernity, by using this machine as a tool of expression, providing a new way of looking at the objects in front of the camera, with a special emphasis on their qualities of line and texture. Photography synthesizes the values of science and art, avoiding the extremes of both materialism and fantasy.[7]

Initially, the rhetoric of straight photography was grounded in aesthetics. Weston and Strand both attempted to defend the artistic potential of photography, arguing that the pictorialists had betrayed that potential by attempting to make photography more like painting. Later many straight photographers turned their attention away from aesthetics to consider the social functions of photography. Strand himself was involved with two related groups, the Photo League and Frontier Films, both of which promoted photography as a valuable tool for social activism.

As Alan Trachtenberg has shown, critics at the time were not sure how to categorize the works of various photographers who did not fit the established definitions of photographic artistry. The cases of Lewis Hine, Walker Evans, and Berenice Abbott were particularly troubling, defying categorization, and falling somewhere between the standard camps of art and journalism. As a way of making sense of their works, in 1938 Beaumont Newhall suggested that one could look at such pictures in two ways, as documents or as photographs. As documents, the pictures performed a valuable social function of recording social problems, providing a visible record with accurate detail. As photographs, the pictures performed a more traditional aesthetic function, rendering their subjects with attention to plastic values like texture and modeling. Trachtenberg's point in his contemporary study is not to endorse Newhall's overly schematic account, but to show how debates about photography were changing over the course of the 1930s. As photographers were abandoning the old debates about photography's status as an art, the term "documentary" became a popular way to make sense of the new approach.[8]

The U.S. government was directly involved in promoting and financing documentary photography and film. For instance, it sponsored films like Pare Lorentz's *The Plow That Broke the Plains* (1936), which was photographed by members of the Photo League, including Strand. Most famously, the government supported photography through the FSA project. In 1935 the Resettlement Administration (later the Farm Security Administration) began an ambitious project, sending out photographers to document the lives of farmers across America. This project produced more than two hundred thousand negatives, including such celebrated images as Dorothea Lange's portraits of migrant workers. The project was shaped by a variety of tensions: between photographers like Lange and government administrators like Roy Stryker, between the desire to create accurate documents and the mandate to produce persuasive propaganda, and between the aesthetic ideals of straight photography and the emerging concept of "documentary photography."[9] From the standpoint of the Hollywood cinematographer, one of the most important sources of tension concerned the status of storytelling in photography. In the twenties, straight photographers were not interested in telling stories. Storytelling was a violation of medium-specificity, recalling the hideous nineteenth-century collages of Henry Peach Robinson, and betraying photography's capacity to capture ephemeral moments in extraordinary descriptive detail. In the thirties, however, with the entire country suffering from the Great Depression, such aesthetic purism seemed like a luxury. In order to engage their audience with the plight of the poor, photographers accepted a new role as public storytellers. Such photographers harnessed the power of evocative moments to construct national narratives about poverty and the need for assistance.

Edward Steichen, perhaps the most influential photographer in the country at the time, celebrated the shift. In a 1938 essay about a major exhibition of FSA photographs, Edward Steichen writes:

> They also found time to produce a series of the most remarkable human documents that were ever rendered in pictures. But the "Art for art's sake" boys in the trade were upset, for these documents told stories and told them with such simple a blunt directness that they made many a citizen wince, and the question of what stop was used, what lens was used, what film was used, what exposure was made, became so completely overshadowed by the story, that even photographers forgot to ask.[10]

In more prosperous times Steichen had been an " 'Art for art's sake' boy" himself, both as a pictorialist and as a straight photographer. Now, confronting devastating national misfortune, he embraced the latest trend, valorizing the photograph as a useful document. Whereas straight photographers like Weston had encouraged photographers to think carefully about their formal choices, such as their decisions regarding f-stop and exposure, Steichen argues that style ought to be subordinate to narrative. The most effective documentary photograph is the one that tells a story.

Many of these trends—from straight photography to documentary photography—would be delivered to the forefront of popular culture by *Life*, which debuted in 1936 as a commercial magazine dedicated to the art of the "photo essay." In 1937 the magazine published several photographs by Weston, describing him as an "intense realist" who used small apertures to produces sharp images with remarkably deep focus.[11] Another issue showed some unusually stylized wide-angle photographs by Photo League members Ralph Steiner and Leo Hurwitz.[12] Berenice Abbott's photographs of New York City appeared in 1938.[13] The magazine also ran several photo essays by their distinguished staff photographers Alfred Eisenstadt and Margaret Bourke-White, including the latter's remarkable 1937 study of southern sharecroppers.[14]

Although *Life* offered many stories about major political events, one of its most important recurring subjects was photography itself. Each issue began with an essay called "Speaking of Pictures," which would usually offer some behind-the-scenes information about a set of photographs—for instance, explaining how a wide-angle lens works, or describing the advantages of photographing with a Leica. *Life*'s interest in the mechanics of photography could even overshadow its interest in politics. One article about a strike shows two images of a man breaking a window. The caption explains, "This picture was taken a split second after the one opposite. The passerby has had time to take just one step. Meanwhile the picket has swung and smacked the pane."[15] The caption shifts attention away from the social causes of the man's actions, instead highlighting the camera's ability to capture ephemeral moments.

The magazine also presents itself in its advertisements as an alternative to glamorous fashion magazines. As the text of one ad rather self-consciously reads:

> Wherever they come from, *Life*'s biographical pictures end up in a union of words and pictures that is neither a detailed timetable of a career, nor a glamour portrait that leaves all the freckles off. It is an impression of a life—the close, intimate impression that that schoolmate or that grandmother who supplied the pictures and their reminiscences might make you feel.[16]

Life certainly contained its share of glamour photos, but it also worked hard to look behind the mask of glamour, running demystifying articles on subjects like "How Starlets Learn to Act."[17] One issue printed a photo of Jean Harlow and pointed out that the image had been rejected by a press agent because it was too unattractive.[18] Instead of glamour, *Life* offered immediacy—the sense that the photograph has captured a moment in time, with the ephemeral nature of the photograph serving as the thumbprint of authenticity.

◉ ◉ ◉

With this context in mind, we can turn to the relationship between photography and cinematography. One of the earliest cinematographers to take note of the changing trends in photography was Karl Struss. Before coming to Hollywood, Struss had owned a portrait studio, invented a popular soft-focus lens, and published photographs in Alfred Stieglitz's journal *Camera Work*. In Hollywood he brought his talents as a portraitist and pictorialist to such revered classics as *Sparrows* and *Sunrise*. In 1934 Struss wrote an article for *American Cinematographer* on "Photographic Modernism and the Cinematographer." Although Struss had based his career on the application of photographic ideals to the cinema, he takes the reverse position here, arguing that "modernist" photography has become too unglamorous to serve as a model for the Hollywood style. Committed to softness, Struss denounces the modernist trend toward sharpness. He writes, "In more than a few instances, realism is carried to the length of exaggeration—especially by the portraitists, who make a fetish of securing exaggeratedly literal renditions of flesh tones and textures, and take greater pains to reveal every possible facial blemish."[19] Struss denounces the trend as "surrealist," but he is not trying to draw a parallel to photographers like Man Ray or to the dreamlike associationist logic characteristic of Salvador Dalí. Instead, he seems to be using the word "surreal" to mean "hyper-real," having too much realism. Unwilling to give photographers the credit for this style, he suggests that photographers are simply imitating the German and Russian classics of the silent cinema. Struss is particularly annoyed by the hyper-realist trend because it goes against his own training as a portrait photographer. He adds, "It is the Cinematographer's duty to minimize these physical imperfections, just as a portrait photographer must strive to please his clientele by presenting his subjects both faithfully and favorably."[20] At heart Struss is a portraitist: when glamour conflicts with realism, then the latter should be sacrificed.

Still, Struss does concede that certain genres could benefit from a harsher style. He offers Raoul Walsh's *What Price Glory* (1926, shot by Barney McGill) and John Ford's *The Lost Patrol* (1934, shot by Harold Wenstrom) as examples. (Apparently the actor Victor McLaglen was so unattractive that he produced a glamour-proof genre all by himself.) The war genre was a suitable home for the hyper-realistic style because the connotations of "hardness" and "toughness" were appropriate for its male-dominated storylines.

Struss makes sense of the new photographic techniques by situating them within the context of the genre/scene conventions. This raises an important question: Is realism suitable for all films, or just a few of them? A few years later *American Cinematographer* addressed this issue in a pair of articles: an interview with Ernst Lubitsch, and a reply from Victor Milner. Lubitsch had just seen Renoir's *The Grand Illusion* (1937) and Duvivier's *Pépé le Moko* (1937), and he was complaining that Hollywood cinematographers were too concerned with technical perfection to create the sense of realism that distinguished those two films.[21] It is a bit disconcerting to find Ernst Lubitsch soapboxing for more realism in Hollywood movies. This was, after all, the man who had a dog bark out the chorus to a Maurice Chevalier song in *The Love Parade* (1929). However, Lubitsch's style had changed over the course of the previous decade, as he dropped the self-conscious touches for a more character-oriented approach.[22]

Lubitsch's former cinematographer, Victor Milner, was president of the American Society of Cinematographers at this time. Perhaps stung by Lubitsch's suggestion that European films were exploring a possibility that American films were not, Milner wrote a highly critical reply. In this response Milner makes two distinct counterarguments. In some sections of the essay, he agrees that all films should be realistic, and he offers explanations as to why they are not. These explanations involve the standard scapegoats of cinematographic rhetoric: producers who do not allow cinematographers to take part in preproduction work, directors who complicate the task of lighting by insisting on moving-camera shots, and stars who prefer to be idealized by cinematography.[23] This rhetorical strategy concedes the point that realism is a worthy goal for all films, while attempting to explain why the goal is not always met. In other sections, Milner is less willing to grant the point that realism is always desirable. Claiming that American audiences prefer lighter fare than European audiences do, Milner argues that most films do not require a realist style. Instead, the realist style works best in prestige productions, like *Fury* (Fritz Lang, 1936, shot by Joseph Ruttenberg), *The Life of Emile Zola* (William Dieterle, 1937, shot by Tony Gaudio), *The Good Earth* (Sidney Franklin, 1937, shot by Karl Freund), and *The Informer* (John Ford, 1935, shot by Joseph August). In response to Lubitsch's criticisms, Milner here explains that the Hollywood cinematographer can and does create realist images when—and only when—the story calls for them. With this second argument, Milner sides with Struss, proposing that realism is just another genre/scene convention.

As we saw in chapter 5, genre/scene conventions follow the logic of associations: cinematographers recommend using a set of techniques while filming a certain genre because they believe that techniques produce associations that can help the audience feel emotions appropriate to the genre in question. For Milner, the realist style is associated with "brutal frankness." "Harshness" is another recurring theme in discussions of the realist style, cited by such cinematographers as Charles Lang and

John Arnold.[24] On a practical level, the primary method of suggesting the traits of frankness or harshness is the elimination of diffusion, particularly lens diffusion. Ever since its introduction to motion picture photography, cinematographers had insisted that diffusion was not supposed to be used in the same way on every film. Instead, cinematographers must vary the level of diffusion in accordance with the type of story being told. By eliminating diffusion, a cinematographer could prepare the audience for *Fury*'s tough-minded look at lynching, or *Emile Zola*'s serious presentation of the historical Dreyfus affair. While these are prestige films, the emphasis on "brutal frankness" and "harshness" can also explain why the realist style is often conflated with the melodramatic style customary for crime films, action films, and horror films. Crime films purport to offer an unflinching look at society's problems (though censors often forced the filmmakers to flinch more than they wanted to), and horror films feature the brutality that was often linked to realism via association.

Although Milner does not cite photography as an influence, it can be useful to contrast his rhetoric with that of straight photography. The discourse of Weston and Strand is pure and absolute. They start with certain assumptions about the nature of photography as a medium, and then they deduce a set of traits that they believe all photographs should have. By contrast, Milner begins with the assumption that a cinematographer needs to command a variety of different styles in order to tell a variety of different stories. Milner does not believe that sharp photography should always be used to take advantage of the medium's essential characteristics, for he believes that photography's most important aspect is its flexibility—its ability to offer the cinematographer a range of stylistic options. We might call this the "toolbox" approach to cinematography. Every stylistic option is a potential tool, waiting to be called upon when the right story comes along. The ideal cinematographer is the one who knows how—and, more important, when—to use each of these tools.

A few years later Milner announced that he was adding another tool to his own toolbox: deep-focus cinematography. By the late thirties, technical developments were allowing cinematographers to use this technique with greater ease. Engineers had fixed many of the problems with arc lights, and Kodak had introduced some new fast film stocks. Some cinematographers took advantage of these developments to reduce light levels on the set. In a 1939 article surveying several cinematographers for their opinions on the new film stocks, Milner is quoted suggesting an alternate approach, proposing that deep-focus cinematography could be used as a new expressive convention:

With the ordinary film we developed a technique of altering the key of our lighting to match the dramatic mood of the action. With the new film we can add to this idea, making the camera more expressive than ever.

For instance, the other day I had a scene in an old-time western saloon and dance hall. It was a big set, bright and full of picturesque action.

Using the new film, that scene could have been lit with half the light I actually used. But instead, I used what would be about a normal lighting for the old film—and compensated by stopping down my lens. That way I gained in depth and crispness in a way that enhanced the mood of the shot.[25]

Here the point is not to suggest that Milner was an innovator in deep-focus cinematography. The technique had been around for a long time, appearing in the silent period (von Stroheim's *Greed*, 1924) and the early sound period (John Ford's *Arrowsmith*, 1931). Milner's advocacy is instead significant because he suggests a thoroughly classical approach to the deep-focus technique. Deep-focus will allow the cinematographer to expand the genre/scene conventions, giving him one more storytelling tool.

Soon Gregg Toland would push the possibilities of deep-focus cinematography to new creative heights. Given that Toland is famous for his deep-focus cinematography, it seems plausible to suppose that he was influenced by the recent developments in photographic technique. For instance, when we look at images like figure 9.1, from *The Grapes of Wrath* (John Ford, 1940), it is not hard to spot the enduring influence of the FSA project. The relatively sharp cinematography encourages us to reflect on the relationship between foreground and background, with the harshness of the landscape explaining the seriousness of the man's face. Like Dorothea Lange, Toland uses his camera as a storytelling device. At the

Figure 9.1 Gregg Toland adopts the style of Depression-era photography for *The Grapes of Wrath* (1940).

same time, we can note how the FSA influence has been incorporated into the style of a Hollywood film. Following the standard conventions for lighting exteriors, Toland has used the sun as a backlight, while employing a cross-frontal key to provide modeling on the man's face. There is an even a tiny point of light sparkling in each of his eyes.

Toland's obsession with deep-focus suggests that he was aware of recent trends in straight photography—after all, Group f/64 was based in California. However, Toland could not afford to indulge in the medium-specificity arguments favored by theorists like Weston and Strand. Weston had justified his own experiments with deep-focus photography by stating that the medium was uniquely capable of capturing details with an extraordinary amount of precision. Toland would need to adopt rhetorical strategies that were more familiar to the ASC.

One option was to follow Milner's lead and argue that his works should be located within the emerging "realist" genre. *Citizen Kane* (Orson Welles, 1941) was quite literally a "ripped-from-the-headlines" picture, with a bold and unflinching tone. Before directing *Kane*, Welles had studied earlier films as models, including John Ford's *Arrowsmith*, which employed deep-focus in the service of a serious, literary drama.[26] In addition to *The Grapes of Wrath*, Toland's contributions to the genre would include the prestigious dramas *The Long Voyage Home* (Ford, 1940) and *The Little Foxes* (William Wyler, 1941). *Ball of Fire* (Howard Hawks, 1941) did not fit the pattern, but Toland could argue that deep-focus was simply an extension of the standard comedic style, which placed a strong emphasis on the overall clarity of the image. In applying the deep-focus style to such films, Toland was partly adhering to the ASC's most basic and enduring principle: the right-mood-for-the-story theory.

However, when we look at Toland's own account of his techniques, we find that realism-as-genre is not his primary justification. Instead, he appeals to a different conception of realism: the illusion of presence. In "Realism for 'Citizen Kane,'" he writes, "The picture should be brought to the screen in such a way that the audience would feel it was looking at reality, rather than merely at a movie."[27] To accomplish this goal, he suggests that cinematography should follow the model of human perception:

> While the human eye is not literally a universal-focus optical instrument, its depth of field is so great, and its focus-changes so completely automatic that for all practical purposes it is a perfect universal focus lens.
>
> In a motion picture, on the other hand, especially in interior scenes filmed at the large apertures commonly employed, there are inevitable limitations.

Even with the 24mm lenses used for extreme wide-angle effects, the depth of field—especially at the focal settings most frequently used in studio work (on the average picture, between 8 and 10 feet for the great majority of shots)—is very small. Of course, audiences have become accustomed to seeing things this way on the screen, with a single point of perfect focus, and everything falling off with greater or less rapidity in front of and behind this particular point. But it is a little note of conventionalized artificiality which bespeaks the mechanics and limitations of photography. And we wished to eliminate these suggestions wherever possible.[28]

If an "invisible observer" were present at the scene represented in figure 9.2, that observer would be shifting his or her attention constantly, moving from Bernstein to Thatcher to Kane and back again. By using a wide-angle lens and a narrow f-stop, Toland creates a deep-focus image that allows the spectator to scan the image in just this way.

Although André Bazin was intrigued by the similarities between deep-focus cinematography and human perception, many people would regard Toland's anatomical account as implausible.[29] In a 1947 article on "Mood in the Motion Picture," Herb Lightman (a future editor of *American Cinematographer*) praises the style of *Citizen Kane* because he believes that its *lack* of realism is perfectly suited to the film's larger-than-life subject: "*Citizen Kane* was notable principally for the number of revolutionary camera techniques which it utilized. Here Toland was told to go all out for effect—to purposely create an unreal perspective of a man's life. He did this with ultra-wide-angle shots, super close-ups, sweeping elevator shots, exaggerated low angles, and radical low-key lighting."[30] Lightman contrasts this bizarre approach with Toland's work in *The Grapes of*

Figure 9.2 Deep-focus cinematography in *Citizen Kane* (1941).

Wrath, where Toland had used a much simpler style to create a documentary atmosphere.

More recent critics have also cast doubt on Toland's claims. Robert Carringer mocks Toland's analysis as "almost comical," questioning how it was possible that Toland and Welles had, without intending to, manufactured one of the flashiest displays of showmanship in Hollywood history.[31] For James Naremore, *Kane* is satisfying because it represents "the limit to which a story could move toward self-conscious 'art' and 'significance' while still remaining within the codes of the studio system."[32] From my point of view, the very fact that Toland's theory is so implausible is itself significant—it suggests the extent to which Toland was straining to justify his technique according to established ideals, such as the illusion of presence and the ethic of invisibility. Toland is pushing against the limits of the Hollywood style, but he is pushing from within.

We could imagine Toland making a different set of rhetorical points, drawing his ideas from outside the ASC. Like Weston, he could have argued that deep-focus photography is valuable because it is *unlike* human perception, allowing us to observe objects with a sense of detail that is unavailable to us in everyday life. Like Strand, he could have insisted that the aesthetic of sharpness is valuable because it expresses the synthesis of art and science, using the precision of a machine to express an aesthetic point of view. Instead, Toland presents his favored technique as the logical culmination of established ASC ideals. Although Bazin would later celebrate deep-focus cinematography as a means of capturing reality in all its ambiguity, Toland presents deep-focus cinematography as a constructed effect. Many of the most celebrated deep-focus shots in the film are actually faked, such as the famous shot of Kane discovering Susan's suicide attempt (fig. 9.3), in which the foreground and background were shot separately. Such fakery would be anathema to a straight photographer, but it is perfectly consistent with Toland's stated desire to give spectators the ability to shift their attention from one story point to another.

Unfortunately for Toland, his peers in the ASC were not all prepared to accept his justifications. Throughout 1941, writers in *American Cinematographer* commented on *Kane*. Almost all the comments were favorable, but they were often qualified by reservations about the broader applicability of Toland's innovations. Surprisingly, Toland did not even win the Academy Award for Best Black-and-White Cinematography that year; the award went to Arthur Miller, for *How Green Was My Valley* (John Ford, 1941). It is not hard to suspect that the Academy was rewarding Miller for using deep-focus cinematography in a more discreet and manageable way.

Figure 9.3 An artificially created deep-focus effect in *Citizen Kane.*

In *The Classical Hollywood Cinema*, David Bordwell argues that such disputes arose from a clash between the values of Art and Craft. The ASC's emphasis on artistry encouraged virtuosity, but its emphasis on craft encouraged invisibility. While Toland may have claimed that deep-focus cinematography was the ultimate in invisible technique, the publicity surrounding his work proved that it was not. By criticizing Toland, the dominant discourse of Craft was keeping Art in check.[33]

I would add that this episode exposes another point of tension: Toland's ideal of realism was at odds with other cinematographic ideals, such as expressivity, glamour, and the illusion of roundness. In a 1942 article, cinematographer Charles G. Clarke offers a detailed critique of Toland's position. First, Clarke doubts if deep-focus is the best analogue for human perception in all contexts: "When you are standing close to a person with whom you are talking—close enough so that you see him in a close-up angle—if your attention is really centered on that person, you cannot be aware of the details of the background."[34] Clarke himself accepts the illusion of presence as a worthy ideal; he simply believes that deep-focus photography fails to meet this ideal.[35] Second, Clarke argues that a certain amount of softness helps to produce the illusion of roundness. According to this theory, a narrow range of focus works like a backlight, separating the foreground from the background. According to Clarke, Toland has needlessly rejected a useful way of suggesting three-dimensional space.[36] Some cinematographers would have added a third criticism: Toland's deep-focus cinematography often required the use of harsh arc lights, thereby violating the conventions of glamour.

On the positive side, Clarke agrees that Toland's deep-focus cinematography works in expressive terms. Here he appeals to the theory of genre that we have been discussing. Crisp, hard images create a suitable mood of harshness and frankness. It is for this reason that Clarke counts *Citizen Kane* itself a success. Offering an unflinching look at a character based on a real-life person, *Kane* easily fits the generic definition of a "realist" film. As a thinly veiled bio-pic, it benefits tremendously from Toland's hard style.

The most interesting aspect of Clarke's argument is his attempt to situate deep-focus photography within a framing narrative of film history. Clarke's historiography mixes the progressive "mechanics-to-artists" narrative with cyclical features. He writes:

> Cinematography, it should be remembered, began its existence as a strictly mechanical process of making a photographic record of scenes and objects in motion. It was not until after the turn of the century that it was discovered that motion picture scenes, properly strung together, could serve as a medium for telling a dramatic story. Since then, the history of the cinema has been a constant search for the best way of combining the inherently accurate mechanical record of the camera with the dramatic and emotional values necessary for true story telling.[37]

Originally, camera operators tried to make everything crisp, because accurate reproduction was their only goal. The art of cinematography began when cinematographers realized that they could create different looks for different kinds of stories. However, Clarke believes that the pendulum went too far in the other direction: during the twenties, films were photographed in the soft style whether the stories called for it or not. Clarke argues (correctly) that crispness returned well before the release of *Citizen Kane*: cinematographers began to use harder images in the early thirties to create harsh moods for somber dramas and gangster films. The return of hardness was a positive development, since it gave cinematographers more choices beyond the homogeneity of softness. Clarke's primary fear is that *Citizen Kane* represents the swing of the pendulum toward another extreme, wherein all films will be photographed in deep-focus. This would be a net loss for the art of cinematography, because art requires options. In order to express the difference between a romance and a crime film, the cinematographer needs two distinct expressive tools and a continuum of choices between the two poles. Clarke welcomes the return of deep-focus to the mix, but only on the condition that it fall in beside soft-focus as an equally useful expressive tool.

While the metaphor of the pendulum swinging back and forth invokes a cyclical model of history, this explanation conflicts with an alternative narrative: that the story of development beyond a stage of mere mechanical reproduction clearly invokes a model of progress. The measurement of progress is not the ever-increasing realism of the cinematic image; it is the ever-expanding set of tools available to the cinematographer, who looks to vary his style to suit the mood of the story. Here we have a historical variant of Milner's "toolbox" theory. As cinematography progresses, the toolbox gets larger. This theory of history helps to explain Clarke's hostility to Toland's work. By invoking the illusion of presence as the guiding ideal, Toland implies that deep-focus cinematography must be employed regardless of the narrative circumstances. Clarke views this argument as a threat. If deep-focus is to be used in every single shot, then the cinematographer will lose many a powerful resource in the modulation of style.

The fact that other cinematographers criticized Toland should not lead us to believe that he was in some sense anticlassical. He is better understood as a mannerist, advocating commonly shared ideals, such as expressivity and the illusion of presence, that were already endorsed by most of his peers in the ASC. The problem is that he tipped the balance of functions. Toland's brand of stylized realism conflicted with other ideals, like glamour and the illusion of roundness. As we have seen throughout this book, many of the best works of Hollywood cinematography are artful compromises, balancing competing functional goals. Toland seemed to be too unwilling to compromise on his approach, and he found himself both inside and outside the Hollywood norms: inside because he was exploring and improving a functional option that was a recognized component of the Hollywood style; outside because his fellow cinematographers worried that such obsessive attention to one technique was bound to upset the delicate balance that a multifunctional style required.

Over the next several years many cinematographers eagerly adopted the technique of deep-focus cinematography. However, they did so by following the toolbox approach, using deep-focus as a new way to achieve previously established functions. For all his criticisms of Toland, Clarke was happy to take advantage of deep-focus when he was shooting the war film *Guadalcanal Diary* (Lewis Seiler, 1943). In figure 9.4, the deep-focus composition lets us keep track of four characters at the same time—an important goal for a genre dedicated to the ethos of the ensemble. The crispness of the image reinforces the genre's masculine connotations, while evoking the sharp, modern style found in the pages of *Life* magazine. The fact that this shot is an exterior is also relevant: Clarke did not have to worry about blinding his actors with an array of arc lights, since he could rely on daylight to provide the general illumination necessary for a

Figure 9.4 Charles Clarke applies deep-focus cinematography to the war genre in *Guadalcanal Diary* (1943).

reduced aperture. Instead, he uses the sun as a top-backlight, while using a high frontal light to provide some simple fill. He even manages to harness and control the sunlight, so that it picks out the two foreground characters, while leaving the two background characters relatively dim.

Clarke modulates the focus over the course of the film, opting for deep-focus in some shots and favoring narrow-focus in others. In figure 9.5, the wide-angle lens, combined with the use of sunlight as a key, would have allowed Clarke to produce a deep-focus image if he had wanted one. However, he chose to narrow the focus just a bit, allowing actor Lloyd Nolan to stand out from the background without giving the background an overly soft, glamorous appearance. The result is an elegant compromise, just sharp enough to fit into the hard-edged genre, and just soft enough to produce the desired illusion of roundness.

As a genre dominated by men, the war film became an ideal genre for experimentation with realist techniques. Still, within the war film we can find a range of different approaches. In *Bataan* (Tay Garnett, 1943), Sid-

Figure 9.5 Shallower focus for a closer shot in *Guadalcanal Diary*.

Figure 9.6 In *Bataan* (1943), the wide shots feature moderately deep-focus cinematography.

Figure 9.7 The closer shots employ shallower focus.

ney Wagner carefully mixes deep-focus and narrow-focus, giving the film some of the expressive hardness of the former, while emphasizing the narrational clarity of the latter. In one scene Sergeant Dane meets the men that have just been placed under his command. Wagner unobtrusively stresses the theme of group solidarity by showing the entire group in moderately deep focus (though the foreground is a bit soft—see fig. 9.6). Dane proceeds to walk down the line, asking each man to introduce himself. The film uses shot/reverse-shot to show Dane's interaction with the two most important characters, located at the beginning and the end of the line. In between these shot/reverse-shot passages, the camera dollies down the line in one unbroken shot, letting us see each man. Here we have an elegant compromise: the unbroken dolly shot serves to emphasize the theme of the group, but the narrow focus and relatively tight framing allows us to appreciate each man as an individual.[38] Meanwhile, Wagner skillfully balances the harsh style of the war film with the glamorous style of a Robert Taylor vehicle. Taylor may look sweaty and dirty, but he has a gentle backlight on his shoulders, and a spark of light in his eyes (fig. 9.7).

By contrast, Rudolph Maté takes the style of *Sahara* (Zoltan Korda, 1943) in a hyper-realist direction (see fig. 9.8). In addition to being a brilliant cinematographer in his own right, Maté was familiar with Toland's style, having worked for producer Samuel Goldwyn during the thirties. Instead of clarifying the story with carefully controlled focus, Maté uses razor-sharp focus to give the story a harsh, brutal mood. Instead of beautifying the shot with puffy clouds in the background, he represents the skies with an austere shade of gray. Instead of glamorizing the star with soft three-point lighting, he creates rugged modeling by using the hard light of the sun as a key-light, with no backlight needed to create separation. The result is a remarkable achievement in cinematographic style: with its sharply focused pictures of sand dunes, the film even manages to evoke the landscape photographs of Edward Weston. Significantly, Maté does not save this style for just a few unusual shots; the entire film has the same mood. It may lack the careful modulation found in a classical film like *Bataan*, but *Sahara* works as an extended exploration of hyper-realist atmosphere.

Does this mean that Karl Struss's nightmare of one-note cinematography has come true? Not necessarily. Although Struss once held that photographic modernism was generally incompatible with the Hollywood style, he conceded that the style might work for a male-oriented action film like *The Lost Patrol*—the very film that inspired *Sahara*'s plot. This is why it would be too extreme to cite *Sahara* as an example of anticlassical modernism. It is more useful to think of it as a mannered variation on the classical style, taking one of the classical ideals (realism defined as hardness) and stretching it to new extremes.

Figure 9.8 Harsh cinematography for *Sahara* (1943). Note the crisp focus and the absence of backlights.

Another genre that would benefit from the sharper style was the crime film. With its hard-hitting tone, this genre had always been linked to the "realist" style. As location shooting became more popular over the course of the decade, deep-focus cinematography became a useful way to situate the characters within the context of the gritty modern city. For instance, in figure 9.9, from *Force of Evil* (Abraham Polonsky, 1948), Toland's old colleague George Barnes uses a wide-angle lens with razor-sharp focus to stage the film's tragic ending against the intricate metal frame of the George Washington Bridge. The image recalls Edward Steichen's famous photograph of the same bridge, but Steichen had been using the wide-angle and deep-focus technique to demonstrate the compositional merits of straight photography. Here, the bridge functions as a metaphor for the modern capitalist city: hard, cold, complex, and vast. Like Clarke, Barnes does not use deep-focus throughout the film. Instead, he saves deep-focus for key dramatic moments like this one, shooting more conventional scenes with softer backgrounds.

Many police procedurals used location photography to produce a style with documentary connotations. Henry Hathaway, who had directed the lushly romantic *Peter Ibbetson* in 1935, turned to the semidocumentary style in a series of films made in the second half of the forties, such as *The House on 92nd Street* (1945) and *13 Rue Madeleine* (1947). A 1947 *American Cinematographer* article interviews cinematographer Norbert Brodine, who photographed both films. By his own account, Brodine names his goal as finding a judicious compromise, taking advantage of the documentary appeal without sacrificing traditional Hollywood values: "We were striving to maintain studio finish, plus a newsreel authenticity."[39] To achieve this effect, he recommends using extensive general illumination in interiors, to ensure a balance between the characters in the foreground

Figure 9.9 Crisp photography for the ending of *Force of Evil* (1948).

and the bright exteriors visible through the background windows. Careful lighting is essential, because a cinematographer could easily lose the illusion of roundness without the aid of studio backlights. In short, Brodine is a classicist, committed to a multifunctional style.

Clearly, the rhetoric of light had embraced a rhetoric of realism—in spite of the fact that "realism" meant so many different things to so many different cinematographers. A final example is James Wong Howe, who represents the culmination of this trend. During the forties, James Wong Howe crafted a distinctive approach to cinematography, combining various aspects of realism. In a 1941 article Howe drew an explicit connection between some of the recent trends in motion picture photography and the sharp style of the magazines:

> Even before Gregg Toland, ASC, came along with his "Citizen Kane," there was a marked tendency in every studio toward crisper definition and greater depth, sometimes accompanied by increased contrast. Better lenses—coated and otherwise—have played their part; so have the snappier contrast of modern emulsions and the improved definition obtainable from fine-grain positive. But to my mind, the biggest factor in this transition has been the change in the public taste. This is directly traceable to the growth in popularity of miniature-camera photography, and to the big picture-magazines like "Life," "Look," and the rest, and such modern photographic magazines as "U.S. Camera," "Popular Photography," and the others. The public has seen the stark realism of the newspicture reporters, and the pictorial strength of the work of the modern miniature-camera photo-illustrators and pictorialists.[40]

Although he had experimented with deep-focus photography in *Transatlantic* (William K. Howard, 1931), Howe had spent the better part of the thirties shooting glamour photography for MGM. *Life*'s edgy style inspired him because it gave him a model for a new kind of cinematography, prioritizing storytelling and realism above glamour and pictorial beauty.

It is one thing for Howe to say that *Life* could be a model; it is another thing for him to put this idea into practice. Howe was committed to realism, but he had to translate that commitment into the lexicon of Hollywood filmmaking, where "realism" could mean an extraordinarily wide variety of things. When we look at Howe's output in the forties and fifties, we find an encyclopedia of realist techniques, as if Howe was attempting to exhaust all of the tools amassed in his toolbox. In *Kings Row* (Sam Wood, 1942), Howe uses Tolandesque deep-focus compositions, with

Figure 9.10 Effect-lighting in the psychological Western *Pursued* (1947).

looming objects in the foreground. In *Air Force* (Howard Hawks, 1943), he employs an unpolished documentary style, partly inspired by wartime newsreels.[41] In *Body and Soul* (Robert Rosson, 1947), Howe crafts unbalanced compositions to create a *Life*-like sense of spontaneity—he even shoots some of the boxing scenes on roller skates, putting the camera in the middle of the action. In *Pursued* (Raoul Walsh, 1947), he exercises special care with effect-lighting, even casting shadows over the actors when the situation demands it (see fig. 9.10). In *The Rose Tattoo* (Daniel Mann, 1955), Howe borrows a strategy from Italian neorealism, producing remarkably unglamorous images of the film's star, Anna Magnani. In his masterpiece, *The Sweet Smell of Success* (Alexander Mackendrick, 1957), Howe combines all of these strategies—while offering sharp location photography of New York City. These accomplishments are even more remarkable when we remember that Howe was originally considered a master of glamour, first celebrated for his ability to photograph the blue eyes of silent film star Mary Miles Minter.[42]

◉ ◉ ◉

The hardening and sharpening of the Hollywood style did not represent a revolutionary realignment of classical lighting techniques, but these changes were not mindless repetitions, either. Instead, we have a series of complex shifts, combining the elements of continuity and change. This brings us back to Bazin's metaphor of the winding river. On one level, the river was still the same—we can find the same conventions, the same functions, the same rhetoric of light. On a deeper level, the river had shifted, as many cinematographers had learned to place a newfound emphasis on the rhetoric of realism.

10 Film Noir and the Limits of Classicism

In the history of Hollywood lighting, no style has received more attention than the film noir. With its darkened corridors and blinking neon lights, the style expands the limits of classical Hollywood storytelling, making it more extravagant and more expressive. As several historians have pointed out, there was no single noir style. There were studio noirs and location noirs; soft-focus noirs and deep-focus noirs; gray noirs and black noirs. To be sure, capturing all of the subtle distinctions and covering all the noir conventions would require a book in itself. Instead, I try here to answer a more general question: What does the heightened expressivity of the noir style tell us about the nature of classical Hollywood lighting? Was noir the antithesis of classicism—or its culmination?

In chapter 7, I proposed a categorical distinction between two different kinds of cinematographers, with these groups coexisting within the membership of the American Society of Cinematographers. A classical cinematographer practices the art of balance, favoring compromise solutions that could accomplish multiple goals. A more mannered cinematographer might reject the multifunctional ideal, sacrificing some norms in pursuit of intensified effects. Upon closer inspection, many film noirs turn out to be surprisingly classical. The lighting is expressive, but the expressivity is expertly balanced with other goals, like glamour or realism. By contrast, the boldest films push the limits of the Hollywood style. Here at the outer edges, film noir is Hollywood mannerism; it takes the widely accepted ideal of expressivity and extends it to new extremes.

Looking at the most influential discussions of noir, we can divide the prevailing critical accounts into two categories: the "expressionist" account and the "revisionist" account. In their influential 1975 article, "Some Visual Motifs of Film Noir," Janey Place and Lowell Peterson argue that noir is consistently antitraditional, reversing the established stylistic norms of the classical cinema. Whereas classical cinema favors shallow-focus and high-key images, the typical film noir is marked by deep-focus cinematography and low-key lighting.[1] Although Place and Peterson do not speculate about the sources of the style, many critics draw a connection between the low-key style of noir and the grotesque distortions familiar to German Expressionism. Paul Schrader points out that many expatriates from Germany and Eastern Europe helped to birth the noir style: the list of directors includes Billy Wilder, Robert Siodmak, and Fritz Lang; among cinematographers, we might cite John Alton, Rudolph Maté, and Karl Freund. Of course, this is not to say that noir is reducible to Expressionism. Rather, the style was marked by a tension between its equally palpable realist and expressionist impulses.[2]

Revisionist historians, such as Barry Salt, Marc Vernet, and Thomas Elsaessar, have problematized the "Expressionist" account, identifying causal flaws in the historical timeline. Barry Salt argues that the word "Expressionist" is habitually misused, pointing out that supposedly Expressionist lighting techniques were actually used in the United States and Denmark before the making of *The Cabinet of Dr. Caligari* (Robert Wiene, 1919).[3] Elsaessar argues that many familiar noir conventions were established in Hollywood during the silent period, only to go "underground" in the thirties, as the horror film and the crime film became minor genres.[4] Although Elsaessar cites the work of Marc Vernet as evidence for his account, a closer look at Vernet reveals that his own revisionism reaches even farther. Rather than suggesting that noir conventions went underground, Vernet posits that night effects remained thoroughly commonplace in classical cinema throughout the thirties and into the forties.[5]

At the level of the conventions, there is a great deal of visual evidence to muster support for the revisionist account. As we have seen in previous chapters, cinematographers adopted effect-lighting and genre/scene conventions from various sources, including the theater and the graphic arts. They proceeded to use those conventions before and after the transition to sound, in both B-films and A-films. Many of the best noir cinematographers were old Hollywood pros who had been around for decades, such as John F. Seitz (*Double Indemnity*, 1944), William Daniels (*The Naked City*, 1948), James Wong Howe (*Body and Soul*, 1947), and George Barnes (*Force of Evil*, 1948). Even Karl Freund had long been assimilated into the Hollywood system. Put simply, most noirs were melodramas, and cinematographers lit them the way they had been lighting melodramas for years: with low-placed key-lights, strong contrasts, and hard cast shadows on the walls.

This does not mean that there was nothing unusual about film noir lighting. It is important to look at noir lighting in context—in the context of the films as a whole, in the context of the history of lighting, and in the context of the culture at large. If we limit our examination to individual shots, we will not necessarily find anything surprising. Figure 10.1, from *The Asphalt Jungle* (John Huston, 1950), certainly appears noir-ish at first glance, with its gloomy shadows and its skillfully executed flashlight effect. This initial impression must be qualified by an awareness that the flashlight effect had been around for a long time—appearing, for instance, in the decidedly un-noir caper film *The Amazing Dr. Clitterhouse* (Anatole Litvak, 1938, shot by Tony Gaudio). Should we therefore conclude that the shot from *The Asphalt Jungle* is noir because Harold Rosson's lighting is so overtly expressive? Or should we conclude that noir is just a dubious category for critics, since most of the techniques were thoroughly conventional? To understand this process of continuity and change, we must be willing to work to interpret the significance of the expressive lighting techniques in each case.

In comparing *The Amazing Dr. Clitterhouse* with *The Asphalt Jungle*, we find that both films are expressive—but we will also find that they have different moods to express. In the earlier film, the somber lighting from below is employed to make the villain seem more menacing. In so doing, the lighting gives the film a more obvious moral structure. The protagonist is a jewel thief, but the film makes him sympathetic by contrasting his crimes with a truly nasty villain. Figure-lighting conventions become useful symbolic shorthand to distinguish between characters along a

Figure 10.1 A flashlight effect in *The Asphalt Jungle* (1950).

spectrum of good and evil. In *The Asphalt Jungle,* the protagonist is another jewel thief with an unusual name (Dr. Reidenschneider), but the film strives to achieve a more complex atmosphere of moral ambivalence that is better suited to the disillusioned atmosphere of the postwar world. When the flashlight picks out the protagonist from the shadows, categories of good and evil are blurred. The policeman pointing the flashlight is not a villain, and Dr. Reidenschneider is not a hero. In the dozen years between these films, the mood of moral clarity has been replaced by a new sense of ambiguity.

In addition to making context-specific interpretations of particular shots, it might be productive to shift the explanation to a different level of abstraction. Ever since they had determined to shed their reputation as "crank-turners," Hollywood cinematographers prized expressivity. The classical cinema is filled with examples of mood lighting; we can find striking examples even in the sugary movies of Shirley Temple. The difference is that expressivity was just one of several *mutually limiting* demands.[6] When the shadows get too dark, they stand to violate a number of other conventions: creating flat planes of blackness that betray the illusion of roundness, obscuring information that is important to the narrative, or producing strong contrasts that destroy the star's glamour. It is not very revealing to say that film noir lighting is expressive. On some level, Hollywood lighting was almost always expressive. The matter meriting deeper inquiry for my study is determining whether film noir lighting is so expressive that it tips the balance of classicism, forcing the cinematographer to sacrifice his other ideals.

An opportune situation for a cinematographer to experiment with new modes of balance came with low-budget films that had relatively little studio oversight. As cinematographer Phil Tannura explains:

> For there are times when, in making a high-budget picture, you may find that the setting or mood of the action aren't such as lend themselves to particularly pictorial effect. You may find a director who is not particularly inclined to cooperate with the photographer, or a star who requires a certain specialized and conservative type of camera-treatment.[7]

By contrast, low-budget films presented the cinematographer a different set of obligations. He had to work fast, but he did not necessarily need to protect the studio's assets by using glamorous photography at all times. Still, we should remember that many films noirs did not fall under the B-film classification. As James Naremore points out, many film noirs occupy a middle ground, somewhere between the "A" and the "B," using stylish production values to compensate for a lack of A-list stars.[8] Cinematographers who worked on such films usually worked on A-films, too—at the very least, they wanted to prove that they could. There are also plenty of films noirs with major stars. In other words, film noir may occupy an unusual position within the system of Hollywood production, but the style still has a place therein. For this reason, we should expect to find some noir cinematographers who favored a more classical approach.

After years of lighting Greta Garbo for MGM, the master classicist William Daniels finally won an Academy Award, for *The Naked City* (Jules Dassin, 1948), a police procedural with noir overtones. Over his long career in Hollywood, Daniels's talent was best expressed through his consummate versatility: during the silent period, he had created sharp, brutal images for Erich von Stroheim. In *The Naked City* he demonstrates his command of the crisply focused, location-based filming that was a vogue for the police procedural drama. Still, for all the location shooting, Daniels has not forgotten all of his old Hollywood tricks. The opening montage presents a series of images from a variety of different locations, as if to suggest that the film will offer a random survey of the city's events in a short period of time. Of course, upon closer examination, the astute viewer will realize that nothing has been left to chance in the montage—many of the characters first introduced here will eventually turn out to be crucial characters in the film's plot. As Edward Dimendberg suggests, the film emphasizes the humanity of the modern city, partly in an effort to compensate for the procedural's relentless logic of science, statistics, and surveillance.[9] The film's visual style works the same way, giving the surface appearance of a newsreel while accomplishing the character-centered functions of traditional lighting. In figure 10.2, from the opening montage, we see two attractive people at a club (Howard Duff and Dorothy Hart). We do not know it yet, but these two characters will be protagonists in the film. Just as he did in *Anna Karenina*, Daniels uses lighting to separate the foreground from the background, backlighting the foreground figures while giving them a strong key. He even moves the key to the left side of the camera, thereby favoring the woman's face. Later in the film, Daniels will craft Dorothy Hart's close-ups with all of the care that he once lavished upon Garbo's portraits. In figure 10.3, Daniels uses a cross-frontal key, ample fill light, gentle backlight, a glimmer of eye-light, and focus so narrow that the background glistens with softness.

Once we see that Daniels is still committed to classical principles, we can identify classical details even in the film's more innovative shots. Fig-

Figure 10.2 The foreground is lit more brightly than the background in this scene from *The Naked City* (1948).

Figure 10.3 *The Naked City*: Conventional three-point lighting.

ure 10.4 is one of the most exhilarating moments in the film, a wide-angle shot showing Jimmy Halloran (Don Taylor) and a colleague ascending over the city in a construction elevator. The crisp lines of the composition are further energized when set against the motion of the active background. The result is a shot that expresses the film's attitude toward the modern city: complex, dynamic, but ultimately ordered. The illumination for this shot is provided by the sun, and its overhead direction is sufficiently unglamorous to give the image an added mark of authenticity. However, it is not so unglamorous that it makes the hero look like a villain. Along with the wide-angle lens, the bright sunlight allows the two men to dominate the composition, arresting our attention in the midst of the highway of moving lines and planes. Here we have a clear example of the multifunctional logic of classicism. Daniels does not have to choose between the demand for narrative clarity and the connotations of realism. As he has proven time and time again, he can have both in the same shot.

The classical style is a *modulating* style. A good cinematographer will adjust the lighting from scene to scene; a great cinematographer will

Figure 10.4 A dynamic shot in an elevator.

adjust the lighting from shot to shot. John F. Seitz was definitely one of the latter. In a crucial scene from *Double Indemnity* (Billy Wilder, 1944), Seitz calibrates his lighting effects to acutely match the changing mood of the story. Phyllis Dietrichson (Barbara Stanwyck) visits the apartment of Walter Neff (Fred MacMurray) and insists that she has no intention of murdering her husband. When the conversation carries into the kitchen, Walter tells her a story about a woman who tried to get away with murder and failed. They move back to the living room, and Phyllis begins to reveal her true intentions. Walter tells her to forget about it, and they embrace. After a brief return to the frame story (Walter is narrating his tale into a dictating machine), we find Walter and Phyllis on the couch, with Walter smoking a cigarette and Phyllis freshening her makeup. Walter pulls Phyllis toward him and tells her that he will help her murder her husband, as long as they do it his way. The scene is superbly written, capturing several shifts in tone: feigned innocence, then a mood of foreboding apprehension in the kitchen, followed by a contrast between Phyllis's murderous intentions and Walter's refusals, all building up to the culminating moment when Walter agrees to commit murder.

Seitz's lighting follows and anticipates each of these tonal shifts with remarkable delicacy. At first, the living room is fairly bright. Phyllis insists that she is innocent, and the lighting does nothing to undercut that assertion (fig. 10.5). Then Walter and Phyllis make their way to the kitchen, where they plausibly but conveniently neglect to turn on the light. This staging decision gives Seitz the pretext he needs to play the next part of the scene in relative darkness—an appropriate shift, given the fact that the mood has turned toward dread (fig. 10.6). Back in the living room, Phyllis sits on the couch next to a lamp. Initially, the diegetic lamp motivates a brighter tonality, just as Phyllis has returned to her initial strategy of denial. A few moments later, Phyllis changes tactics again, confessing that she is thinking about murder after all. Here, Seitz and Wilder confront a style-meets-story problem: the atmosphere should be shifting toward darkness again, but Phyllis is still sitting next to a lamp. They devise an elegant solution. Walter moves from the chair to the couch, forcing Phyllis to turn her head away from the lamp. Now her face is in shadow for the moment when her murderous intentions are revealed (fig. 10.7). In the final part of the scene, Seitz turns off one of the lamps, another suggestive hint that the two characters have had sex in the elided portion of the story. The effect-lighting is skillfully executed, with Phyllis seemingly lit by a table lamp and Walter in the shadows (fig. 10.8). Now it is Walter who is associated with the darkness, and Seitz places shadowy character-lighting on Walter's face as he discusses his own plans for the murder (fig. 10.9). As a whole, the sequence is a superb display of the art of modulation. A lesser cinematographer might have cast the entire scene in shadows, following a routine convention for melodramas. Yet, true to his skilled reputation, Seitz intensifies the expressive effects by tweaking his lighting in almost every shot, in perfect harmony with the moment-by-moment progression of the scene.

Remarkably, Seitz manages to respect several other conventions, even as he fine-tunes the genre/scene conventions to a point of maximum precision. Figure 10.5 establishes an internal norm for the scene, lighting Phyllis's face from a high frontal position to smooth her features, and outlining her hair with a traditional backlight. Figure 10.6 is much darker, but it is actually a variation on the same strategy: a high frontal key with a backlight. Seitz has dimmed the key and placed a flag over the lower half, creating the necessary tone of darkness without sacrificing the advantages of the original schema: the key-light placement still smoothes Phyllis's skin, and the backlight still serves to separate her from the background. A bright eye-light allows us to appreciate the determination in her gaze. With this image in mind, the lighting in figure 10.9 takes on more significance. Instead of using a dim, low-contrast effect, Seitz opts for a more contrasty scheme, using a cross-frontal key with relatively weak fill and no backlight. The result follows the norms of masculine portraiture, etching the character on Walter's face at a moment where he is getting delirious with a (misguided) sense of power.

As we have seen, many classical films solved the problem of conflicting conventions by adopting a "gendered expressivity" strategy. In films like *Cleopatra* and *Only Angels Have Wings*, cinematographers balanced conflicting demands by glamorizing the female star while creating an

Figure 10.5 *Double Indemnity* (1944): Relatively bright lighting as Phyllis (Barbara Stanwyck) denies being interested in murder.

Figure 10.6 Darker tonalities when the subject of murder returns.

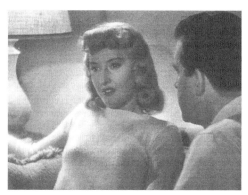

Figure 10.7 Discussing murder more seriously, Phyllis turns away from the light

expressive arc that mirrored the dramatic arc of the male protagonist. The woman's face is illuminated, allowing it to become the site of unin-hibited emotional expression. The man's face is more stoic, registering fewer expressive gestures, but the lighting style compensates by symbol-izing his emotions for him. One way to understand some of the new lighting trends in the forties is to think of them as a response to larger changes in the culture's understanding of emotional expression. On the

Figure 10.8 Later, one lamp is turned off, and Walter (Fred MacMurray) is in the shadow.

Figure 10.9 Walter's face is dark when he agrees to commit the murder.

one hand, the cumulative impact of the Depression and the war encouraged a culture of stoicism, a steady resolve in the face of daunting challenges. On the other hand, those same challenges produced cultural expressions of anxiety, paranoia, and dread—emotions that were emphasized by the increasingly popular culture of psychoanalysis. These conflicting approaches to emotional expression could be represented in gendered terms. While some films noir celebrate the toughness of the hard-boiled hero, the coldness of the femme fatale is designed to be jarring, as it departs from the previously established feminine norm. Equally disturbing is the emotionally overwrought neuroticism of certain noir men, which conflicts with prevailing norms of masculinity.[10]

Occasionally, a film noir will celebrate a relatively hard-boiled woman. Place and Peterson astutely offer *High Sierra* (Raoul Walsh, 1941) as an example. Although Roy (Humphrey Bogart) initially falls for the seemingly sweet Velma (Joan Leslie), it is clear that he is a much better match for Marie, the tougher character played by Ida Lupino. In one remarkable scene, Roy examines the bruises on Marie's face. Even though Lupino was the top-billed star, cinematographer Tony Gaudio lights her face with

Figure 10.10 Ida Lupino lit from below in *High Sierra* (1941).

a low, hard key-light, imitating the effect of a table lamp (fig. 10.10). It is important to be clear about why this is such an unusual scene. The salient point is not that Gaudio has sacrificed the narrative to create an extravagant, moody effect. Quite the contrary: the striking thing is that Gaudio is committed to storytelling here—so committed that he is willing to sacrifice glamour in the interest of appropriate characterization. When we recall that James Wong Howe was allegedly fired from MGM for making a similar choice a few years before, we can appreciate all the more how the film noir allows cinematographers to create a different kind of balance.

Another solution to the problem of gendered figure-lighting conventions was to use varying degrees of glamour for different characters. In *Ministry of Fear* (Fritz Lang, 1944), veteran cinematographer Henry Sharp uses lighting to differentiate between the femme fatale and the good-girl female protagonist. In figure 10.11, the wholesome Carla (Marjorie Reynolds) is lit in conventional fashion, with a bright key-light illuminating the majority of her face. Significantly, her lighting does not change much over the course of the film. No matter how grave the impending danger, she will always be lit with bright frontal key. By contrast, Mrs. Bellane, the femme fatale (Hillary Brooke), receives much more expressive variation, even though she is a less important character. In the séance scene, Sharp lights her from below, using a standard technique usually reserved for male villains (fig. 10.12). Later, Sharp lights her for glamour, using a front-cross key to cast a conspicuously beautiful pattern on her face (fig. 10.13). As a noir detective might say, "She looks good—too good." Whereas Carla is associated with unobtrusive beauty, the femme fatale becomes an image of false glamour.

Figure 10.11 In *Ministry of Fear* (1944), Marjorie Reynolds is lit in a simple but flattering style.

Figure 10.12 *Ministry of Fear*: The femme fatale lit from below.

Figure 10.13 The femme fatale with a pattern of shadows on her face.

Murder, My Sweet (Edward Dmytryk, 1944) also establishes a simplistic split between the two major female characters, but it does occasionally strive for a more ambivalent atmosphere. In the club scene, Philip Marlowe (Dick Powell) begins to suspect that good-girl Ann Grayle (Anne Shirley) is not as good as she seems. Although she will turn out to be good in the end, the scene expresses film noir's characteristically paranoid atti-

Figure 10.14 Patterns on the face of Anne Shirley in *Murder, My Sweet* (1944).

tude toward women: any woman can be a femme fatale. Fittingly, cinematographer Harry Wild lights Ann from the side, leaving half of her face in shadow. The illuminated side of her face is darkened by a carefully motivated crisscross pattern (fig. 10.14). The result is still glamorous, but it departs from the typical feminine ideal, which uses bright frontal lighting to make the woman's face the site of unambiguous emotional expression.

Marlowe himself is morally unambiguous—as Raymond Chandler would say, he is the man who walks the mean streets, but who is not himself mean. However, the film still expresses deep anxieties about masculinity by representing him as a man who cannot control his own emotions. In a bizarre dream sequence, the rational detective Marlowe must experience the irrational, running through a series of doors while trying to escape from a syringe-wielding doctor. When his surreal escape attempt ends in futility, the film shows a close-up of Marlowe's tormented face (fig. 10.15). For most of the film, Wild has used unobtrusive three-point lighting for Marlowe. Here he uses a more unusual lighting scheme to bring out all the tortured wrinkles on Marlowe's face, with a cross-light

Figure 10.15 Strong modeling emphasizes the anxiety on Dick Powell's face.

molding the right side of his face, and a kicker modeling the left. The asymmetrical eye-lights increase the sense of psychological confusion.

◎ ◎ ◎

These examples move away from the center, which represents classicism, and inch slightly toward the edges of mannerism. Throughout this book I have maintained that the Hollywood lighting style is defined by various tensions, such as the tensions between the ASC and the studios, and the tensions between glamour and expressivity. Classical cinematographers *resolve* these tensions via their subtle compromises and their gradual shifts in tone. Mannerist cinematographers *expose* these tensions, revealing that the problem of conflicting conventions cannot always be solved. Perhaps no other cinematographer exposes those tensions more clearly than John Alton, the ultimate noir cinematographer—and, arguably, the ultimate mannerist.

Born in a region of Europe that is now part of Hungary, Alton emigrated to the United States in 1919. After working as a cinematographer in New York, Los Angeles, and Paris, Alton spent most of the thirties shooting films in Buenos Aires for the emerging Argentine film industry. He returned to Hollywood in 1939, and he spent the next several years shooting B-films for various studios. Alton was a member of the ASC, but he struggled with John Arnold, the powerful president of the organization, and he soon quit the society in disgust. In the late forties Alton shot a series of brilliant film noirs with Anthony Mann. Working at the low-budget Eagle-Lion, they collaborated on the following pictures: *T-Men* (Mann, 1947), *Raw Deal* (Mann, 1948), *He Walked by Night* (1948, with direction credited to Alfred Werker), and *The Black Book* (Mann, 1949; alternatively titled *Reign of Terror*). The last one was a truly bizarre concoction, mixing the film noir style with a plot about the French Revolution, often interpreted as an allegory about anti-Communist paranoia. Alton and Mann also collaborated on *Border Incident* (1949), released by MGM. Throughout this period, though he generally refrained from discussions within *American Cinematographer*, Alton contributed regularly to the journal *International Photographer*. In 1949 he published a book on cinematography, entitled *Painting with Light*. In the next decade Alton worked primarily at MGM, where he again met opposition from his old ASC nemesis, John Arnold, now the head of MGM's camera department. Fortunately, major directors like Vincente Minnelli and Richard Brooks insisted on working with him, and Alton welcomed this work as the opportunity to experiment with color cinematography. In 1960, still at the height of his powers, Alton retired.[11]

It is easy to read the story of Alton as the story of a daring experimentalist who broke all the rules: He formed his personal aesthetic while working outside of the Hollywood system, in Europe and Argentina. He perfected that aesthetic while working at the margins of Hollywood, in low-budget crime films. Although he was accepted into the ASC, he quit the organization. He eventually reached the peak of the profession anyway, but even his Oscar could not prevent him from having to struggle with John Arnold. Unwilling to be subordinate to the dominant rules any longer, he decided to quit on his own terms.

There is certainly some truth to this story: Alton was an experimental cinematographer who tried to push the limits of the Hollywood style. However, we also need to recognize the other side of the story: as much as he struggled with the Hollywood system, Alton was a product of it. As James Naremore has remarked, Alton's book *Painting with Light* is not the noir manifesto we might suppose it to be; it is a textbook of studio lighting techniques, written from the point of view of a man who was perhaps "more of a Sternbergian aesthete than a tough guy."[12] Throughout *Painting with Light*, Alton employs the pronoun "we." This suggests that the book was intended to be a summary of Hollywood lighting conventions as they were practiced throughout the industry. Alton may be the book's narrator, but his advice is practiced by a collective body of artisans. Alton consistently emphasizes the aesthetic justifications for the conventions he describes, but this is not unusual: cinematographers had always thought of Hollywood lighting in aesthetic terms. Even the book's title, *Painting with Light*, had long ago become a cliché of cinematographic discourse. Nineteen years earlier, Victor Milner had used the same title for his contribution to the 1930 volume of the *Cinematographic Annual*.

Alton's general account of Hollywood lighting is consistent with Hollywood norms. He explicitly argues that lighting should be multifunctional:

> The purpose of illumination is twofold, that is, for quantity and for quality. In lighting for *quantity*, we light for exposure, to make certain that a sufficient amount of light reaches every corner of the set, and that it is properly balanced, in order that no part of the film shall be underexposed or overexposed. In lighting for *quality*, we strive to bring out the following values:
>
> 1. Orientation—to enable the audience to see where the story is taking place
> 2. Mood or feeling (season of year and time of day)

3. Pictorial beauty, aesthetic pleasure

4. Depth, perspective, third-dimensional illusion.[13]

At a minimum, there must be enough lighting to produce an appropriate exposure. Once this minimum level has been met, the cinematographer can attend to more aesthetic concerns, like mood, pictorial beauty, and the illusion of roundness. Even the most classical of Hollywood cinematographers would not quibble with this account.

Nor would any Hollywood cinematographer be surprised by most of Alton's specific recommendations. Discussing exterior landscapes, Alton advises, "To get the necessary depth, foreground pieces have to be darker than the background; if the sun happens to hit them, keep it off by properly shading them with goboes or flags. There is no prettier picture than the setting sun with a silhouette in the foreground" (123). Alton may be the master of the gritty film noir, but he also knows how to shoot pretty exteriors, recommending the use of staid techniques, like the *repoussoir* and the silhouette, that had been popularized by Maurice Tourneur back in the silent period.

Similarly, Alton is in full command of Hollywood's glamour conventions, advising his readers to use a high frontal key with ample fill light and backlight when shooting a feminine close-up (94). There is even an astonishing chapter called "Day and Night, Ladies, Watch Your Light." Here Alton argues that even far away from the studio sets, average women should pay attention to the way they are lit in everyday life. For instance, he argues that better lighting in the workplace will produce greater professional efficiency: "Better light would make the girls look prettier, consequently feel happier. It is easier to get work out of happy people than out of gripers" (176). Alton goes on to offer detailed advice about how women should be lit when doing specific activities, like eating dinner or listening to the radio.

Even Alton's discussion of "Mystery Lighting" grows out of a long-standing Hollywood convention: the rule advising the cinematographer to use effect-lighting to produce the shadowy style associated with the melodrama. Alton sees no contradiction in the idea that lighting can be expressive and realistic at the same time. He argues: "People are getting tired of the chocolate-coated photography of yesterday. They have had enough of it. In the latest films there seems to be a tendency to go realistic. . . . There is no better opportunity for such realistic lighting than in mystery illumination" (45). Alton does not present mystery lighting as an exercise in unmotivated Expressionism. Rather, mystery lighting uses thoroughly conventional effects (cigarette lighters, flashing neon lights, hanging lamps, and so on) to produce an appropriate set of emotional associations. Like Walter Lundin and Victor Milner before him, Alton is a firm believer in an old Hollywood adage: pick the right mood for the story.

This does not mean that Alton has simply adopted Hollywood's rhetoric of light without revision. Although Alton's book is surprisingly conventional, his films are stunningly original. If Garmes is a pictorialist, and Howe is a hyper-realist, then Alton is an expressivist, willing to sacrifice various functions in the pursuit of an intensely atmospheric mood. For instance, many cinematographers would ensure a basic level of exposure by using a large number of overhead lamps for general illumination. In addition to meeting a minimum standard of pictorial quality, the technique helps to create an illusion of roundness, by eliminating flat pools of shadow. Alton prefers to use lights on floor stands, which allows him to make adjustments more quickly. By reducing the level of general illumination, he produces darker shadows. Such a strategy has benefits and drawbacks. On screen, the primary benefit is increased expressivity. If shadows create mood, then deep shadows will create intense moods. The problem is that this expressive benefit comes at an inevitable cost, producing impenetrable pools of blackness that threaten to flatten the illusion of roundness.

T-Men is a superb example of Alton's mannerist style. Working with real locations and no-name actors, Alton uses a minimum of lights to produce a maximum amount of mood. In figure 10.16, Moxie (Charles McGraw) emerges from a shadow, just before he commits a murder. The idea of using shadows for a murder scene is not new; the idea of creating this effect with just one light is. There is no fill light to add detail to the shadows, and no backlight to separate Moxie from the background. He simply emerges from a flat plane of blackness. A more classical cinematographer would prefer a compromise solution here, with just enough shadow to create mood, and just enough light to preserve detail.

Alton rejects the compromise solution because he is working with an idiosyncratic theory of mood. As Alton writes, "Where there is no light, one cannot see; and when one cannot see, his imagination starts to run wild. He begins to suspect that something is about to happen. In the dark there is mystery" (44). In other words, the lack of detail is precisely the strategy that produces the most mysterious mood. Because we cannot see the space clearly, we begin to suspect the worst. The theory produces an interesting corollary: Alton can create a sense of mystery by using light tonali-

Figure 10.16 A simple but dramatic lighting scheme in *T-Men*
(1947).

ties just as easily as he can by using dark tonalities. In the steam room
scenes, Alton takes this idea to its logical conclusion, giving us a murder
scene where the bulk of the space is represented in whites and grays (fig.
10.17). The steam performs the same function as the shadow, rendering
the space illegible and inciting the viewer's imagination.

Although Alton scoffs at "chocolate-coated" cinematography, he re-
spects the Hollywood norm that women should look glamorous at all
times. Following a strategy used by other noir cinematographers, Alton
lights the good girl differently from the femme fatale. Tony's loving wife
Mary (June Lockhart) receives a simple frontal key, with a glowing eye-
light (fig. 10.18). Evangeline, a photographer involved in the counterfeiting
ring, is introduced with a more unusual arrangement featuring a crisscross pattern on the side of her face (fig. 10.19). The effect is glamorous—
with all the connotations of falseness that "glamour" implies. Of course,
there is nothing unusual about a film switching from expressive effect-

Figure 10.17 A steam effect
obscures the background.

lighting to glamorous figure-lighting when a woman appears onscreen. The unusual thing is that Alton makes no effort to produce a smooth transition. A more classical cinematographer would shift from a moderately expressive style to a moderately glamorous style, attempting to get the benefits of both strategies without calling undue attention to the shift. In 1936 Alton's arch-nemesis John Arnold had written that a great cinematographer was one who could make such transitions seamlessly. By contrast, Alton is never moderate; his expressive style is so intensely expressive that the rare glamorous moments stand out as aberrations.

Alton is committed to multifunctionalism as a general ideal, but his experimentation with expressive effects tips the balance away from the classical norm. While we might criticize Alton for ignoring the nuanced solutions that highlight the work of a cinematographer like William Daniels, we can celebrate him for daring to produce images that no one else in Hollywood would imagine. Alton can do more with one light than most cinematographers can do with twenty. As a final example of the virtues of Alton's expressive style, consider the scene of Tony's murder.

Figure 10.18 Frontal lighting for June Lockhart, playing the wife of an undercover agent.

Figure 10.19 Patterned lighting for a femme fatale.

As Tony (Alfred Ryder) tries to convince the members of the counterfeiting gang that O'Brien (Dennis O'Keefe) knew nothing about Tony's true identity as a Treasury agent, the gang members eye O'Brien suspiciously. One thug, Shiv (John Wengraf), is lit from below. So far, Alton has done nothing new—in his book, Alton himself admits that the convention of lighting criminals from below dates back to the silent period. It is the next few moments that stand out. Ignoring the rules of continuity, Alton lights O'Brien with a hard light in a high frontal position (fig. 10.20). The hardness gives crisp definition to the lines on his face, while the frontality allows us to witness his tense emotional response, fearing the impending death of his friend and worrying that his own identity as a T-man might also be revealed. As Tony dies, O'Brien lowers his head in grief, the shadow of his hat slowly inching down until his entire face is enshrouded in darkness (fig. 10.21). It is an extraordinary effect—all accomplished with the careful placement of a simple key-light.

While it is important to note that this mannered solution violates the classical norm of putting light on the protagonist's face, it is equally

Figure 10.20 As he watches his partner die, the protagonist lowers his head . . .

Figure 10.21

. . . and his hat casts a shadow over his face.

important to realize that Alton is amplifying a different classical norm. In films like *Mr. Deeds Goes to Town* and *Only Angels Have Wings*, Joseph Walker had used expressive shadows to underline the emotional experience of male characters who were unwilling to express their own emotions directly. Here John Alton expresses the grief of a character who is exerting great force to keep his grief hidden inside. The difference is that Walker had employed this strategy within a set of classical constraints, always ensuring that his boss Harry Cohn could see every detail of the picture. Alton goes beyond those constraints, making his shadows even more important than the actors. Perhaps it is fitting that figure 10.21 is the last image in this book. In mannerist fashion, Alton has taken the classical concept of "gendered expressivity" to its logical conclusion—creating the evocative image of a man whose expression is almost literally invisible.

◉ ◉ ◉

We can now return to the question introduced at the beginning of this chapter: Was noir the antithesis of classicism—or its culmination? Decades before, John F. Seitz had argued that the narrative film was the perfect forum for beautiful, expressive cinematography. It is not unreasonable to see Seitz's work in *Double Indemnity* as the culmination of that idea. In comparison to Seitz, John Alton is much more willing to sacrifice the principle of balance in the pursuit of heightened expressivity. Still, we cannot say that *T-Men* is the antithesis of classicism. In its own way, it, too, is the culmination of a long-standing goal—a goal that was present at the founding of the ASC. Like Seitz, Alton was what so many cinematographers aspired to be: an artist.

Conclusion: Epilogue

A logical end date for my historical account can be placed around 1950, when the careers of many great Hollywood cinematographers came to an end. Some cinematographers passed away: Joseph August (1947), Gregg Toland (1948—at the age of 44), Joseph Valentine (1949), Tony Gaudio (1951), and George Barnes (1953). Other cinematographers retired, including major figures like Arthur Edeson, Victor Milner, Arthur Miller, Sol Polito, and Joseph Walker. Several veterans switched to a new medium, lending their expertise to television. Karl Freund, who used three-camera shooting on *I Love Lucy*, is the most famous example, but he was soon joined by other old pros like Norbert Brodine, Robert de Grasse, and Henry Sharp. Even the cinematographers who kept working in the movies found themselves in new situations: by 1950, William Daniels had left MGM, James Wong Howe had left Warners, Ernest Palmer had left Fox, and Karl Struss had left Paramount. The Supreme Court's 1948 Paramount decision hastened the process, encouraging studios to cut costs by eliminating some of their long-term staff contracts. A few cinematographers would maintain their long-standing relationships with production studios (George Folsey stayed at MGM for another decade, and Leon Shamroy dominated cinematography at Fox until the late sixties), but these were exceptional cases. The studio system, in the classic sense of the term, was coming to end.

There are other reasons why the fifties would mark a turning point in the ongoing annals of Hollywood cinematography. Led by Fox, the studios began employing widescreen formats to compete with television and other popular leisure activities. Kodak introduced a new monopack film stock, which proved to be more convenient than Technicolor—though it was not quite as glorious. With these and other technical advancements, location shooting became more feasible, and some producers opted to shoot in exotic locations for added production value. On a more general level, we might say that declining audiences forced Hollywood filmmakers to make fewer, more spectacular films: altering the balance of the classical style by placing ever more weight on pictorial display.

Of course, the conventions of Hollywood lighting did not disappear. Many of the lighting conventions discussed in this book are still used today. Returning to Bazin's metaphor, we can say that the river had carved out a fresh equilibrium profile, featuring a new balance of functions. With looser ties to the studios, a cinematographer might take more risks with glamour lighting. With a desire to create a contrast with television, another cinematographer might experiment with colorful pictorialism. With the need to maintain a reputation as a free agent moving from job to job, a third cinematographer might look for opportunities to employ an eye-catching expressive style and define a personal "look."

Some would argue that these post-Paramount changes were changes for the better. For some tastes, the classical Hollywood style is too neutral, too bland, too redundant. The James Wong Howe of *The Sweet Smell of Success* (Alexander Mackendrick, 1957) seems bolder than the James Wong Howe of *Algiers* (John Cromwell, 1938); the color cinematography of *Written on the Wind* (Douglas Sirk, 1956) seems more vivid than the color cinematography of *The Garden of Allah*; the shadowy lighting of late film noir seems more expressive than the shadowy lighting of *After the Thin Man*. These are valid comparisons, but I hope to have shown that the lighting of the classic studio period was richer than it may at first appear. The style of Hollywood lighting was certainly conventional, but the conventions did not reduce the complexity of the style; if anything, they increased it. Every scene presented a new problem, requiring the nimble cinematographer to balance the demands of genre with the demands of the scene, or the needs of the star with the needs of the role.

In conclusion, we can think of the rhetoric of light as a set of ideas that almost overlap, and a set of conventions that almost interlock. These "almosts" created an endless series of puzzles for the Hollywood cinematographer. The art of Hollywood cinematography was the art of solving those puzzles.

Notes

The following abbreviations are used throughout these notes:

AC = *American Cinematographer*
JSMPE = *Journal of the Society of Motion Picture Engineers*
TSMPE = *Transactions of the Society of Motion Picture Engineers*

Introduction: The Rhetoric of Light

1. During this period all Hollywood cinematographers were men. For that reason I refer to cinematographers as "he" and "him" throughout.

2. Here my argument both draws on and departs from studies in the semiotics tradition, such as Sharon Russell, *Semiotics and Lighting: A Study of Six Modern French Cameramen* (Ann Arbor: UMI Research Press, 1981); and Mike Cormack, *Ideology and Cinematography in Hollywood, 1930–39* (New York: St. Martin's Press, 1994). Although I stress the existence of lighting conventions, I do not insist that they are arbitrary. Rather, they are guided by functional concerns.

3. Phil Tannura, "What Do We Mean When We Talk About Effect Lighting?," *AC* 22 (March 1942): 25.

4. David Bordwell, Janet Staiger, and Kristin Thompson, *The Classical Hollywood Cinema: Film Style and Mode of Production to 1960* (New York: Columbia University Press, 1985). See, in particular, ch. 1, "An Excessively Obvious Cinema," 3–11; and ch. 7, "The Bounds of Difference," 70–84.

5. For instance, discussing the various choices made in the making of any film, Robert Ray has written, "The American Cinema's formal paradigm, however, developed precisely as a means for concealing these choices. Its ability to do so turned on this style's most basic procedure: the systematic subordination of every element to the interests of a movie's narrative" (p. 32). This argument gives invisibility a higher priority than narrativization. By comparison,

Noël Burch places extra emphasis on Hollywood's ability to offer the spectator the experience of seeing the fictional world from the position of an ideal
observer. Burch calls this experience the "transcendental voyage." Whereas
the spectator of early cinema (in Burch's terms, the "primitive mode of representation") is aware that she is looking at images on display, the spectator of
Hollywood cinema (part of Burch's "institutional mode of representation")
experiences the fictional world from within, as the camera moves the spectator from one ideal viewing location to another. See Robert B. Ray, *A Certain
Tendency of the Hollywood Cinema, 1930–1980* (Princeton: Princeton University Press, 1985), 32–55; and Noël Burch, *Life to Those Shadows*, ed. and trans.
Ben Brewster (London: British Film Institute, 1990), 202–233.

6.   Richard Maltby, *Hollywood Cinema: An Introduction* (Cambridge: Blackwell,
1995), 30–35. I discuss the disagreement between Bordwell and Maltby in
more detail in Patrick Keating, "Emotional Curves and Linear Narratives,"
Velvet Light Trap 58 (fall 2006): 4–15. For other examples of scholars offering
criticisms of the "classical" model, see Rick Altman, "Dickens, Griffith, and
Film Theory Today," in *Classical Hollywood Narrative: The Paradigm Wars*, ed.
Jane Gaines (Durham: Duke University Press, 1992), 9–48; Elizabeth Cowie,
"Storytelling: Classical Hollywood Cinema and Classical Narrative," in *Con-
temporary Hollywood Cinema*, ed. Steve Neale and Murray Smith (New York:
Routledge, 1998), 178–190; Donald Crafton, "Pie and Chase," in *Classical Hol-
lywood Comedy*, ed. Kristine Brunovska Karnick and Henry Jenkins (New
York: Routledge, 1994), 106–119; Dirk Eitzen, "Comedy and Classicism," in
Film Theory and Philosophy, ed. Richard Allen and Murray Smith (New York:
Oxford University Press, 1999), 393–411; Miriam Bratu Hansen, "The Mass
Production of the Senses: Classical Cinema as Vernacular Modernism," in
Reinventing Film Studies, ed. Christine Gledhill and Linda Williams (London:
Arnold, 2000), 332–350; and Linda Williams, "Melodrama Revised," in *Refig-
uring American Film Genres: History and Theory*, ed. Nick Browne (Berkeley:
University of California Press, 1998), 42–88.

7.   A study that employs a similar methodology is Scott Higgins, *Harnessing the
Technicolor Rainbow: Color Design in the 1930s* (Austin: University of Texas
Press, 2007). Barry Salt also studies the relationship between stylistic norms
and institutions, but his primary focus is on the role of technology in the
development of film style; Barry Salt, *Film Style and Technology: History and
Analysis*, 2nd ed. (London: Starword, 1992).

8.   See Cormack, *Ideology and Cinematography*, esp. ch. 2.

9.   James Lastra, *Sound Technology and the American Cinema: Perception, Representation, Modernity* (New York: Columbia University Press, 2000). See in
particular chs. 5 and 6, 154–215.

10.   David Bordwell has proposed a "target" model of scholarship, with the individual film at the center and the culture at large on the outer edge, with various mediating layers in between; for one version of this proposal, see *Poetics
of Cinema* (New York: Routledge, 2008), esp. ch. 1. For a similar idea, see Pe-

ter Gay's "hourglass" model, in *Art and Act: On Causes in History—Manet, Gropius, Mondrian* (New York: Harper and Row, 1976), 11.

11. Joseph Ruttenberg, quoted in the microform *Joseph Ruttenberg: An American Film Institute Seminar on His Work* (Glen Rock, N.J.: Microfilming Corporation of America, 1977), 67.

1. Mechanics or Artists?

1. William E. Fildew, "Trials of the Cameraman," *Moving Picture World* (July 21, 1917): 391.

2. Anonymous, "An Artist in Camera Angles," *Moving Picture World* (April 23, 1927): 714.

3. Quoted in Silas Snyder, "Motion Picture Cameramen's Organizations in America," *International Photographer* 7.9 (Oct. 1935): 3. See also H. Lyman Broening, "How It All Happened: A Brief Review of the Beginnings of the American Society of Cinematographers," *AC* 2 (Nov. 1, 1921): 13. For a more recent history, see Robert S. Birchard, "The Founding Fathers," *AC* 85 (Aug. 2004): 54–65; and Birchard, "Shaping Cinematography's 'Magazine of Record,'" *AC* 85 (Aug. 2004): 66–75.

4. Quoted in Anonymous, "American Cinematographer Twenty Years Old," *AC* 21 (Nov. 1940): 507.

5. Anonymous, "The American Cinematographer," *AC* 3 (July 1, 1922): 8. With this increase in ambition, there was an increase in the magazine's size. This first issue had only four pages. By the middle of 1922, the standard length had become 28 pages.

6. For these four article titles, see the following issues of *American Cinematographer*: Nov. 1, 1921; Nov. 1924; July 1926; and Aug. 1928.

7. As far as I know, the author of these articles was never identified. One possibility is that they were written by Karl Brown, the author of *Adventures with D. W. Griffith*. Formerly an assistant to Billy Bitzer, Brown was an associate editor for the magazine. A short biographical sketch of Brown in one issue mentions that he "is a clever satirical writer"; see *AC* 2 (Feb. 1, 1922): 48.

8. "Jimmy the Assistant: The Golden Gift of Gab," *AC* 7 (Jan. 1927): 28.

9. "Jimmy the Assistant: Wages and Salaries," *AC* 9 (May 1928): 40. In a different context, cultural historian Warren Susman has written about the rise of a new professional class during this period, noting that the distinction between salaries and wages helped define the class; see Warren Susman, *Culture as History: The Transformation of American Society in the Twentieth Century* (Washington, D.C.: Smithsonian Institution Press, 2003), xxi.

10. "Jimmy the Assistant: The Golden Gift of Gab," 28–29.

11. William Marshall, "Art and Commercialism," *AC* 4 (Dec. 1923): 5.

12. "Jimmy the Assistant: Directors," *AC* 2 (Nov. 1, 1921): 12.

13. "Jimmy the Assistant: On Art and Business," *AC* 3 (Sept. 1922): 17.

14. Joseph A. Dubray, "Evolution," *AC* 3 (June 1, 1922): 26.

15. See Anonymous, "The Qualitative Picture: The Influence of the French School of Picture Making," *Moving Picture World* (June 25, 1910): 1089–1090; Lux Graphicus, "On the Screen," *Moving Picture World* (July 30, 1910): 241; Anonymous, "Pictorialism and the Picture: A Biograph Pastoral," *Moving Picture World* (Sept. 10, 1910): 566–567; Anonymous, "The Usual Thing," *Moving Picture World* (Oct. 1, 1910): 738. Some of these articles are discussed in Charlie Kiel, *Early American Cinema in Transition: Story, Style, and Filmmaking, 1907–1913* (Madison: University of Wisconsin Press, 2001).

16. See Anonymous, "For the Cameraman," *Moving Picture World* (July 10, 1915): 282; Carl Louis Gregory, "Motion Picture Photography," *Moving Picture World* (July 10, 1915): 283; Carl Louis Gregory, "Motion Picture Photography," *Moving Picture World* (March 25, 1916): 2006.

17. See, for instance, Wid Gunning, "Oh, I Think This Will Answer," *Wid's Films and Film Folk* 1.7 (Oct. 21, 1915): 1.

18. Anonymous, "The Usual Thing: Second Article," *Moving Picture World* (Oct. 8, 1910): 798–799. Many film scholars have examined the film industry's attempt to use self-consciously artistic appeals to attract middle-class audiences in this period, though they do not always agree in their conclusions. For three different approaches, see Tom Gunning, *D. W. Griffith and the Origins of American Narrative Film: The Early Years at Biograph* (Urbana: University of Illinois Press, 1991); Sumiko Higashi, *Cecil B. DeMille and American Culture: The Silent Era* (Berkeley: University of California Press, 1994); and Lary May, *Screening Out the Past: The Birth of Mass Culture and the Motion Picture Industry* (Chicago: University of Chicago Press, 1980).

19. Anonymous, "The Usual Thing," 738.

20. For a classic discussion of the relationship between taste and cultural hierarchies in American culture, see Lawrence W. Levine, *Highbrow/Lowbrow: The Emergence of Cultural Hierarchy in America* (Cambridge, Mass.: Harvard University Press, 1988).

21. Foster Goss, "The Editor's Lens: A Rose by Any Other Name, But—," *AC* 5 (Feb. 1925): 10–11.

22. Foster Goss, "The Editor's Lens: 'Culture' and Cinematography," *AC* 6 (March 1926): 12.

23. Goss, "'Culture,'" 12.

24. Joseph A. Dubray, "Art vs. Commercialism," *AC* 2 (Feb. 1, 1922): 4.

25. Foster Goss, "The Editor's Lens," *AC* 4 (Aug. 1924): 10.

26. Daniel B. Clark, "What It Takes to Be a Cinematographer," *AC* 6 (Aug. 1926): 11.

27. Philip E. Rosen, "Cinematography as the Director's Aid," *AC* 3 (Oct. 1922): 4.

28. Harry Fischbeck, "Lighting and the Cameraman," *TSMPE* (Nov. 1926): 144. Although this passage did not appear in the pages of *American Cinematographer*, it is consistent with my argument that the ASC had shaped the larger discourse, encouraging a widespread awareness of the artistic possibilities of lighting.

29. Virgil E. Miller, "The Cinema Triangle," *AC* 7 (Jan. 1927): 4.

30. Miller, "Cinema Triangle," 4.

31. David Bordwell, Janet Staiger, and Kristin Thompson, *The Classical Holly-wood Cinema: Film Style and Mode of Production to 1960* (New York: Columbia University Press, 1985), 345.

32. Dubray, "Art," 6.

33. Rosen, "Cinematography," 4.

34. Foster Goss, "The Editor's Corner: Subservient Art?," *AC* 7 (Oct. 1926): 5.

2. From the Portrait to the Close-Up

1. This chapter draws on my research for Patrick Keating, "From the Portrait to the Close-Up: Gender and Technology in Still Photography and Hollywood Cinematography," *Cinema Journal* 45.3 (spring 2006): 90–108.

2. Arthur Hammond, "Home Portraiture," *American Photography* (March 1917): 130.

3. Hammond, "Home Portraiture," 130.

4. Henry Peach Robinson, *The Studio, and What to Do in It* (1891; New York: Arno Press, 1973), 42.

5. Early critics had argued that photography was too mechanical to be considered an art. For instance, an 1857 article proposes that photography does not have the selectivity of a true art. Instead, the camera is good at offering impartial evidence of the facts, as in the manner of an "unreasoning machine"; see Lady Elizabeth Eastlake, "Photography," *Classic Essays on Photography*, ed. Alan Trachtenberg (New Haven: Leete's Island Books, 1980), 66.

6. T. W. Kilmer, "Artificial Lighting in Portraiture," *American Annual of Photography* (1918): 68.

7. Kilmer, "Artificial Lighting," 70.

8. Note: consulting this British journal seems like a fair strategy, given that Charles Rosher, one of the most important portraitists-turned-cinematographers, was British. See Geo. F. Greenfield, "An Efficient Method of Working the Enclosed Arc," *British Journal of Photography* (Jan. 31, 1913): 79–81.

9. George R. Henderson, "Portraiture with the Open Arc," *British Journal of Photography* (Jan. 28, 1916): 53–54.

10. Arthur Hammond, "Portraiture: Lighting," *American Photography* (Nov. 1920): 606.

11. Hammond, "Portraiture: Lighting," 607, 610.

12. Antony Guest, *Art and the Camera* (1907; New York: Arno Press, 1973), 134.

13. Paul L. Anderson, *The Fine Art of Photography* (1919; New York: Arno Press, 1973), 237.

14. Anderson, *Fine Art of Photography*, 238–239.

15. Anderson, *Fine Art of Photography*, 281–282.

16. Hammond, "Portraiture: Lighting," 618.

17. Guest, *Art and the Camera*, 147.

18. Anonymous, untitled article, *Photographic Review* 24.1 (Jan. 1918): 8.

19. See Anonymous, "The Focusing Screen," *Photographic Review* 23.2 (Feb. 1917): 28.

20. For a more detailed account of De Meyer, see Anne Ehrenkranz, *A Singular Elegance* (San Francisco: Chronicle Books, 1994), 13–49.

21. Anonymous, "Our Illustrations," *Studio Light* (June 1917): 6.

22. Edward Steichen, *A Life in Photography* (New York: Doubleday, 1963).

23. For more information on this transition, see Barry Salt, *Film Style and Technology: History and Analysis*, 2nd ed. (London: Starword, 1992), 115–126.

24. In the following discussion, most of my examples will come from relatively close shots, such as medium shots or close-ups. There are two reasons for this—one practical, one substantive. The practical reason is that it makes analysis easier, since we can look more closely at the figure-lighting in close shots. The substantive reason is that the figure-lighting conventions were quite simply more important in the close shots than they were in wide shots. In a wide shot, a cinematographer had to integrate the demands of figure-lighting with other priorities, such as the needs of the set. In a closer shot, the lighting of the human face was usually the most crucial task, and figure-lighting was given the highest priority.

25. See Salt, *Film Style and Technology*, 130–131, 161; also see Kristin Thompson's discussion of the "soft style" in David Bordwell, Janet Staiger, and Kristin Thompson, *The Classical Hollywood Cinema: Film Style and Mode of Production to 1960* (New York: Columbia University Press, 1985), 287–293.

26. Anonymous, "John Leezer," *AC* 3 (Feb. 1, 1922): 40.

27. Fred J. Balshofer and Arthur C. Miller, *One Reel a Week* (Berkeley: University of California Press, 1967), 150.

28. G. W. Bitzer, *Billy Bitzer: His Story* (New York: Farrar, Straus, and Giroux, 1973), 84.

29. Karl Brown, *Adventures with D .W. Griffith*, ed. Kevin Brownlow (New York: Da Capo Press, 1976), 99.

30. I discuss the early years of backlighting in more detail in Patrick Keating, "The Birth of Backlighting in the Classical Cinema," *Aura* 6.2 (2000): 45–56.

31. Other reasons backlight is more noticeable with female stars: first, men are more likely to have very short haircuts, which do not pick up the backlight as well; and second, female stars are more likely to have blonde hair, which reflects the light more easily (at least, after the conversion to panchromatic film).

32. Salt, *Film Style and Technology*, 117.

33. See Kevin Brownlow, *The Parade's Gone By* (Berkeley: University of California Press, 1968), 227.

34. Salt, *Film Style and Technology*, 154.

35. Rex Ingram, "Directing the Picture," in *Hollywood Directors, 1914–1940*, ed. Richard Koszarski (New York: Oxford University Press, 1976), 89.

36. See illustration 56 in Steichen's *A Life in Photography*.

37. See Miriam Hansen, *Babel and Babylon: Spectatorship in American Silent Film* (Cambridge, Mass.: Harvard University Press, 1991), 245–294; and Gaylyn Studlar, *This Mad Masquerade: Stardom and Masculinity in the Jazz Age* (New York: Columbia University Press, 1996), 150–198.

3. The Drama of Light

1. Fabrice Revault d'Allonnes, *La lumière au cinéma* (Paris: Cahiers du cinéma, 1991), 7. My translation.

2. Peter Baxter also discusses the influence of the theater in his essay "On the History and Ideology of Film Lighting," *Screen* 16.3 (autumn 1975): 83–106. Baxter quotes a later publication by Pichel.

3. Irving Pichel, "Stage Machinery and Lighting Equipment," *Theatre Arts Magazine* 4.2 (April 1920): 143–144.

4. For a general survey of this history, see Gösta Bergman, *Lighting in the Theatre* (Totowa, N.J.: Rowman and Littlefield, 1977).

5. Theodore Fuchs, *Stage Lighting* (Boston: Little, Brown, and Co., 1929), 4.

6. Samuel Selden and Hunton Sellman, *Stage Scenery and Lighting: A Handbook for Non-Professionals* (New York: F. S. Crofts and Co., 1930), 208–209.

7. Louis Erhardt and S. R. McCandless, "The Procedure for Lighting a Production," *Our Theatre Today: A Composite Handbook on the Art, Craft, and Management of the Contemporary Theatre*, ed. Herschel L. Bricker (New York: Samuel French, 1936), 291.

8. Fuchs, *Stage Lighting*, 441. In the original, Fuchs italicizes the word "simulated," to draw a contrast between simulated effects and projected effects— such as the projection of a lightning bolt during a storm scene.

9. Fuchs, *Stage Lighting*, 438.

10. Edward Gordon Craig, *On the Art of the Theatre* (1911; London: Heinemann, 1957), 27.

11. Fuchs, *Stage Lighting*, 10.

12. Sheldon Cheney, "The Failure of the American Producer," *The Theatre* 20.161 (July 1914): 6–10.

13. Cheney, "Failure of the American Producer," 10.

14. David Belasco, *The Theatre Through Its Stage Door* (New York: Harper and Brothers, 1919), 162. For more information on Belasco's lighting techniques, see Louis Hartmann, *Theatre Lighting* (1930; New York: Drama Book Specialists, 1970).

15. Belasco, *Theatre Through Its Stage Door*, 183.

16. Bayard Veiller, *Within the Law: A Melodrama in Four Acts* (1912; New York: Samuel French, 1917), 96–97.

17. John Willard, *The Cat and the Canary: A Melodrama in Three Acts* (1921; New York: Samuel French, 1927), 41. Willard's original copyright was in 1921, and

the play was first staged in 1922. It is possible that these stage directions were added for the 1927 edition, but it is not implausible that the 1922 staging had some lighting effects of this kind. In any case, there are several other melodramas that could be cited as examples. See, for example: Robert L. Parkinson, *The Irish Detective: A Melodrama in Four Acts* (Philadelphia: Penn Publishing Company, 1906), 48–49; W. S. Maugham, *The Explorer: A Melodrama in Four Acts* (London: William Heinemann, 1912), 38; Roi Cooper Megrue, *Under Cover* (1914; New York: Samuel French, 1918), 80; Elmer Reizenstein, *On Trial: A Dramatic Composition in Four Acts* (1914; New York: Samuel French, 1919), 23–24; and William Edwin Barry, *The Jade God: A Mystery Play in Three Acts* (1928; New York: Samuel French, 1930), 35–36.

18. For information on the lighting style of Antoine, see Jean Clothia, *André Antoine* (New York: Cambridge University Press, 1991), 60–79.

19. See J. L. Styan, *Max Reinhardt* (New York: Cambridge University Press, 1982), and Scott Eyman, *Ernst Lubitsch: Laughter in Paradise* (Baltimore: Johns Hopkins University Press, 2000), 28–40.

20. See Scott MacQueen, "Roland West," *Between Action and Cut: Five American Directors,* ed. Frank Thompson (Metuchen: Scarecrow Press, 1985), 105–162.

21. See Matthew Paul Solomon, "Stage Magic and the Silent Cinema" (diss., University of California–Los Angeles, 2001).

22. When the film was rereleased in 1918, the villain was renamed Haka Arakau. See Sumiko Higashi, *Cecil B. DeMille and American Culture: The Silent Era* (Berkeley: University of California Press, 1994), 101.

23. Lea Jacobs, "Belasco, DeMille, and the Development of Lasky Lighting," *Film History* 5 (1993): 404–418.

24. Anonymous, "Lighting to Photoplay Is Like Music to Drama Declares Cecil B. DeMille," *Motography* (Jan. 29, 1916): 249.

25. For a discussion of Orientalism in *The Cheat,* see Higashi, *Cecil B. DeMille and American Culture,* 100–112.

26. Cecil B. DeMille, "Motion Picture Directing," *TSMPE* 12.34 (1928): 300. This is from a speech that DeMille originally gave in 1927.

27. Tom Gunning discusses Griffith's cultural ambitions in *D. W. Griffith and the Origins of American Narrative Film: The Early Years at Biograph* (Champaign: University of Illinois Press, 1991).

28. Both pictures can be found in Marianne Fulton Margolis, ed., *Camera Work: A Pictorial Guide* (New York: Dover, 1978).

29. For a critical introduction to the theory of pictorialism, see Joel Eisinger, *Trace and Transformation: American Criticism of Photography in the Modernist Period* (Albuquerque: University of New Mexico Press, 1999).

30. See art historian Nancy Forgione's essay " 'The Shadow Only': Shadow and Silhouette in Late Nineteenth-Century Paris," *Art Bulletin* 81.3 (Sept. 1999): 494.

31. James Wong Howe, "Course in Direction, Lecture Four" (July 9, 1943): 4. The notes for this lecture are in folder 88 of the James Wong Howe papers at the Margaret Herrick Library in Los Angeles.

32. Geo. R. Henderson, "Effects by Artificial Lighting," *British Journal of Photography* (Dec. 13, 1912): 953.

33. Charles Higham, *Hollywood Cameramen: Sources of Light* (Bloomington: Indiana University Press, 1970), 66.

34. Bert Glennon, "Film Psychology and The 10 Commandments," *AC* 5 (Aug. 1924): 5.

35. Glennon, "Film Psychology," 6.

36. Steve Neale, "Melo Talk: On the Meaning and Use of the Term 'Melodrama' in the American Trade Press," *Velvet Light Trap* 32 (fall 1993): 66–89.

37. Barry Salt, "From Caligari to Who?," *Sight and Sound* 48 (spring 1979): 119–123.

38. John Alton, *Painting with Light* (1949; Berkeley: University of California Press, 1995), 54–55.

39. I have not been able to locate an early edition of this play. However, the 1959 edition is filled with stage directions involving shadowy lighting effects. Given the already mentioned tradition of expressive lighting in the melodramatic theater, it is likely that the original staging in 1920 included at least some of these effects. See Mary Roberts Rinehart and Avery Hopwood, *The Bat: A Play of Mystery in Three Acts* (1920; New York: Samuel French, 1959).

40. See Karl Brown, *Adventures with D .W. Griffith*, ed. Kevin Brownlow (New York: Da Capo Press, 1976), 52.

41. Both of these illustrations can be found in Gustave Doré and Blanchard Jerrold, *London: A Pilgrimage* (1872; New York: Dover, 1970). *The Bull's-Eye* is between 144 and 145; *Opium Smoking* is between 146 and 147.

42. See *The National Police Gazette* 48 (23 Dec. 1893): 5. For other examples from this surprisingly morbid magazine, see the following issues: 23 (21 Sept. 1867): 1151; 33 (21 Dec. 1878): 65; 48 (23 Sept. 1893): 9; 49 (19 May 1894): 16; 54 (18 Feb. 1899): 12; and 55 (27 Oct. 1900): 8.

43. There are several examples in a 1902 edition of the *Collected Works of Edgar Allan Poe* (New York: Lamb, 1902). For instance, see: vol. 3, p. 39; vol. 4, p. 172; vol. 7, p. 164; vol. 9, p. 190; and vol. 10, p. 186. For more examples, see *Works of Edgar Allan Poe*, ed. Edmund Clarence Stedman and George Edward Woodberry (New York: Charles Scribner's Sons, 1914); in particular: vol. 1, p. 16; and vol. 3, p. 36.

44. Goss, "The Editor's Lens," *AC* 7 (Aug. 1926): 5. Goss is referring to the following article: "Camera Dynamics," *Film Daily* (27 June 1927): 25–27. The *Film Daily* article is unsigned, but it was probably written by Maurice Kann, the editor. Kann had praised Dupont's *Variety* several times in the previous months.

45. See the profile of Warrenton in *AC* 2 (Feb. 1, 1922).

46. Fred J. Balshofer and Arthur C. Miller, *One Reel a Week* (Berkeley: University of California Press, 1967), 123.

47. Virgil E. Miller, "The Cinema Triangle," *AC* 7 (Jan. 1927): 4.

48. Antonio Gaudio, "Difficulties of Screen Photography: Natural Lighting and the Necessary Screen Illusion," *Moving Picture World* (21 July 1917): 393.

4. Organizing the Image

1.   Cecil B. DeMille, *The Autobiography of Cecil B. DeMille*, ed. Donald Hayne (Englewood Cliffs, N.J.: Prentice-Hall, 1959), 144. Thanks to Ben Brewster for alerting me to this passage.

2.   Scott Eyman, ed., *Five American Cinematographers* (Metuchen, N.J.: Scarecrow Press, 1987), 65.

3.   Alvin Wyckoff, "Motion Picture Lighting," *Annals of the American Academy of Political and Social Sciences* (November 1926): 63–64.

4.   See Barry Salt, *Film Style and Technology: History and Analysis*, 2nd ed. (London: Starword, 1992), 76. Salt points out that writers often used the term "Cooper-Hewitt" to refer to any mercury vapor lamp, whether it was manufactured by the Cooper-Hewitt company or not (p. 321).

5.   See Adolphe Appia, *Music and the Art of the Theatre*, trans. Robert W. Corrigan and Mary Douglas Dirks (Coral Gables: University of Miami Press, 1962), 74.

6.   For instance, Samuel Selden and Hunton Sellman cite Appia when they propose a distinction between general and specific illumination; see Selden and Sellman, *Stage Scenery and Lighting: A Handbook for Non-Professionals* (New York: F. S. Crofts and Co., 1930), 207–208.

7.   Douglas E. Brown, "Cine Light," *TSMPE* 16 (May 7–10, 1923): 42. As noted in the previous chapter, Karl Brown (unrelated, as far as I know) also mentions Doré as an influence on cinematic composition. Doré uses the darkened foreground technique in many prints; see, for instance, "The River Bank," in Gustave Doré and Blanchard Jerrold, *London: A Pilgrimage* (1872; New York: Dover, 1970).

8.   Wilfred Buckland, "The Scenic Side of the Photodrama," *Moving Picture World* (July 21, 1917): 374–375.

9.   Charles G. Clarke, *Charles Clarke's Professional Cinematography* (1964; Hollywood: American Society of Cinematographers, 2002), 107.

10. See David Bordwell, Janet Staiger, and Kristin Thompson, *The Classical Hollywood Style: Film Style and Mode of Production to 1960* (New York: Columbia University Press, 1985), 287–293.

11. Stephen S. Norton, George Meehan, L. Guy Wilky, and Jackson J. Rose, "Uses and Abuses of Gauze," *AC* 4 (July 1923): 12.

12. Hal Mohr, quoted in the microform *Hal Mohr: An American Film Institute Seminar on His Work* (Glen Rock, N.J.: Microfilming Corporation of America, 1977), 30–31.

13. John F. Seitz, "Can a School Teach Cinematography?," *AC* 4 (July 1923): 5.

14. Seitz, "School," 5.

15. There are several important articles on this topic. See, in particular, Tom Gunning, "The Cinema of Attractions: Early Film, Its Spectator, and the Avant-Garde," in *Early Cinema: Space, Frame, Narrative*, ed. Thomas Elsaessar (London: British Film Institute, 1990), 56–62; and Gunning, "Response to 'Pie and Chase,'" in *Classical Hollywood Comedy*, ed. Kristine Brunovska Karnick and Henry Jenkins (New York: Routledge, 1995), 120–122. Also, Steve

Neale and Frank Krutnik, *Popular Film and Television Comedy* (New York: Routledge, 1990), esp. ch. 6; Donald Crafton, "Pie and Chase: Gag, Spectacle, and Narrative in Slapstick Comedy," in *Classical Hollywood Comedy*, ed. Karnick and Jenkins, 106–119.

16. Walter Lundin, "Drama Treatment Enters Comedy Photography," *AC* 5 (June 1924): 9.

17. Lundin, "Drama Treatment," 10.

18. L. Guy Wilky, "Behind the Camera for William de Mille," *AC* 6 (Feb. 1926): 7.

19. See Robert B. Ray, *A Certain Tendency of the Hollywood Cinema, 1930–1980* (Princeton: Princeton University Press, 1985), 32–55; and Noël Burch, *Life to Those Shadows*, ed. and trans. Ben Brewster (London: British Film Institute, 1990), 202–233.

5. Inventing the Observer

1. A. Lindsley Lane, "The Omniscient Eye," *AC* 15 (March 1935): 95. Lane also wrote "Rhythmic Flow—Mental and Visual," *AC* 15 (April 1935): 138–139, 151–152; and "Cinematographer Plays Leading Part in Group of Creative Minds," *AC* 15 (Feb. 1935): 48–49, 58.

2. David Bordwell, Janet Staiger, and Kristin Thompson, *The Classical Hollywood Cinema: Film Style and Mode of Production to 1960* (New York: Columbia University Press, 1985), 30; and James Lastra, *Sound Technology and the American Cinema: Perception, Representation, Modernity* (New York: Columbia University Press, 2000), 194.

3. Arthur Hopkins, "Our Unreasonable Theatre," *Theatre Arts Magazine* 2.2 (Feb. 1918): 79–84.

4. Thomas Puttfarken discusses the ideal of illusionism throughout his book *The Discovery of Pictorial Composition: Theories of Visual Order in Painting, 1400–1800* (New Haven: Yale University Press, 2000). Interestingly, while he argues that illusionism was an important goal for Renaissance painting, he suggests that it was somewhat less important for other other traditions, such as French Academic painting.

5. For an authoritative history of the transition, see Donald Crafton, *The Talkies: American Cinema's Transition to Sound, 1926–1931* (Berkeley: University of California Press, 1999). Crafton draws on the ideas of a number of scholars, including Rick Altman, David Bordwell, and Douglas Gomery.

6. The next several pages benefit from the following two works: Barry Salt, *Film Style and Technology: History and Analysis*, 2nd ed. (London: Starword, 1992), 179–194; and Bordwell, Staiger, and Thompson, *Classical Hollywood Cinema*—in particular, Bordwell's chapters on the introduction of sound, 298–308.

7. Ben Brewster, in conversation. See also Salt, *Film Style and Technology*, 181; and Kevin Brownlow, *The Parade's Gone By . . .* (Berkeley: University of California Press, 1968), 212. However, note that some cinematographers preferred standardized processing. Referring to silent-era processing, Karl

Struss complained, "Before, often the lab man would develop the negative to his own taste, neutralizing what I'd been trying to do. They can't play with it any more" (in Scott Eyman, ed., *Five American Cinematographers* [Metuchen, N.J.: Scarecrow Press, 1987], 21).

8. Charles Clarke, *Highlights and Shadows: The Memoirs of a Hollywood Cameraman*, ed. Anthony Slide (Metuchen, N.J.: Scarecrow Press, 1989), 34.

9. See Bordwell's chapter on this topic in *Classical Hollywood Cinema*, 294–297.

10. See the editorial in *International Photographer* 1 (March 1929): 5.

11. See the page of anecdotes and jokes in *International Photographer* 1 (June 1929): 10.

12. See Mike Nielsen and Gene Mailes, *Hollywood's Other Blacklist: Union Struggles in the Studio System* (London: British Film Institute, 1995), 9–14. See also Gerald Horne, *Class Struggle in Hollywood, 1930–1950: Moguls, Mobsters, Stars, Reds, and Trade Unionists* (Austin: University of Texas Press, 2001), 43–44. Both books argue persuasively that the failure to support the 1933 strike was a terrible setback for all Hollywood unions.

13. Joseph A. Dubray, "Evolution," *AC* 3 (June 1, 1922): 26.

14. A. S. Howell and J. A. Dubray, "Some Practical Aspects of and Recommendations on Wide Film Standards," *JSMPE* 14 (Jan. 1930): 60.

15. Ben Schlanger, "On the Relation Between the Shape of the Projected Picture, the Areas of Vision, and Cinematographic Technic," *JSMPE* 24 (May 1935): 405.

16. John F. Seitz, "Introduction," in *Cinematographic Annual*, vol. 1, ed. Hal Hall (Hollywood: American Society of Cinematographers, 1930), 18.

17. Thanks to Ben Singer for pointing out the potential reference to the Magnascope system.

18. Alfred N. Goldsmith, "Problems in Motion Picture Engineering," *JSMPE* 23 (Dec. 1934): 350.

19. See Lastra, *Sound Technology*, ch. 6.

20. Joseph Walker and Juanita Walker, *The Light on Her Face* (Hollywood: ASC Press, 1984), 171. Hal Mohr makes a similar complaint; see Leonard Maltin, *The Art of the Cinematographer*, enl. ed. (New York: Dover, 1978), 82.

21. Bert Glennon, "Cinematography and the Talkies," *AC* 10 (Feb. 1930): 7.

22. Seitz, "Introduction," 13.

23. Charles B. Lang Jr., "The Purpose and Practice of Diffusion," *AC* 14 (Sept. 1933): 171.

24. F. W. Murnau, "Films of the Future," in *Hollywood Directors, 1914–1940*, ed. Richard Koszarski (New York: Oxford University Press, 1976), 219. This anthology also contains an interesting article by Paul Fejos, entitled "Illusion on the Screen." Fejos, like Murnau, was an artistic filmmaker from Europe known for his bold camera movements. In this article Fejos argues that the production of an illusion of presence is one of the highest goals that an artist can strive for. This supports my hypothesis that it is sometimes advisable to split apart the concepts of illusionism, invisibility, and classicism—in both of these examples, a flamboyant stylist invokes presence as a guiding ideal.

25. Victor Milner, "Let's Stop Abusing Camera Movement," *AC* 15 (Feb. 1935): 46–47, 58–59. The article includes quotations from James Wong Howe, Tony Gaudio, Charles Lang, George Folsey, and Frank B. Good. See also Hal Hall, "Cinematographers and Directors Meet," *AC* 13 (Aug. 1932): 10, 47. For a scholarly discussion, see William O. Huie, "Style and Technology in Trouble in Paradise: Evidence of a Technicians' Lobby?," *Journal of Film and Video* 39 (spring 1987): 37–51.

26. Quoted in Charles Higham, *Hollywood Cameramen: Sources of Light* (Bloomington: Indiana University Press, 1970), 72. Similarly, Stanley Cortez once said that cinematographers "are dramatists to a very high degree, as a writer is"; see the microform *Stanley Cortez: An American Film Institute Seminar on His Work* (Glen Rock, N.J.: Microfilming Corporation of America, 1977), 25.

27. James Wong Howe, "Lighting," in *Cinematographic Annual*, vol. 2, ed. Hal Hall (Hollywood: American Society of Cinematographers, 1931), 47.

28. L. Owens Huggins, "The 'Language' of Tone," *AC* 14 (Feb. 1934): 398. For the other articles in Huggins's series, see: "The 'Language of Line' in Photography," *AC* 14 (Jan. 1934): 354; "The 'Language' of Design," *AC* 14 (March 1934): 440; and "The Language of Color," *AC* 15 (July 1934): 108.

29. Charles G. Clarke, *Charles Clarke's Professional Cinematography* (1964; Hollywood: American Society of Cinematographers, 2002), 97.

30. Slavko Vorkapich, "Cinematics," in *Cinematographic Annual*, vol. 1, ed. Hal Hall (Hollywood: American Society of Cinematographers, 1930), 33.

31. Many philosophers would question the theory of pictorial perception that underlies this account. For instance, it might be argued that it is not possible to see the two-dimensional pattern and the three-dimensional diegetic world at the same time. For a thorough discussion of these issues, see Richard Allen, "Looking at Motion Pictures," in *Film Theory and Philosophy*, ed. Richard Allen and Murray Smith (New York: Oxford University Press, 1997), 76–94.

32. Victor Milner, "'Miscasting' the Cinematographer," *AC* 13 (Feb. 1933): 13.

33. Victor Milner, "Creating Moods with Light," *AC* 15 (Jan. 1935): 7.

34. Long after the studio period had ended, cinematographers from this era would continue to cite mood as a guiding ideal. For instance, consider two examples from Scott Eyman's book of interviews, *Five American Cinematographers* (Metuchen, N.J.: Scarecrow Press, 1987. Referring to *Gaslight* (George Cukor, 1944), Joseph Ruttenberg says, "That film had the most valuable thing a film can have: mood" (44). In response to a question about what he looks for in a script, aside from a good story, James Wong Howe replies, "Opportunities for dramatic lighting, and mood" (88).

6. Conventions and Functions

1. Charles Clarke, *Highlights and Shadows: The Memoirs of a Hollywood Cameraman*, ed. Anthony Slide (Metuchen, N.J.: Scarecrow Press, 1989), 127.

2. William Stull, "Amateur Movie Maker," *AC* 11 (Oct. 1930): 34.

3. George W. Hesse, "Shadows," *AC* 13 (June 1932): 37.

4. Hal Wallis, memo to Sam Bischoff, Aug. 18, 1939, in the *Roaring Twenties* files in the Warner Bros. Archives at USC.

5. Charles G. Clarke, *Charles Clarke's Professional Cinematography* (1964; Hollywood: American Society of Cinematographers, 2002), 119.

6. Charles G. Clarke, "Reflectors for Exterior Photography," *AC* 12 (March 1932): 34, 37. Clarke is primarily writing for amateur readers, but he implies that most of these tools are also available to the professional.

7. Vladimir Nilsen, *The Cinema as a Graphic Art*, trans. Stephen Garry (New York: Hill and Wang, 1959), 177–179.

8. Barry Salt, *Film Style and Technology: History and Analysis*, 2nd ed. (London: Starword, 1992), 229. See also James Wong Howe's discussion of the two studio styles in Scott Eyman, ed., *Five American Cinematographers* (Metuchen, N.J.: Scarecrow Press, 1987), 76.

9. Phil Tannura, "What Do We Mean When We Talk About Effect Lighting?," *AC* 22 (March 1942): 25.

10. Tony Gaudio, "Precision Lighting," *AC* 18 (July 1937): 278. On the use of small spotlights in the late thirties, see Salt, *Film Style and Technology*, 202.

11. Tony Gaudio, "Difficulties of Screen Photography: Natural Photography and the Necessary Screen Illusion," *Moving Picture World* (July 21, 1917): 392–393.

12. Gaudio, "Precision Lighting," 288.

13. Arthur C. Miller, "Putting Naturalness Into Modern Interior Lighting," *AC* 21 (March 1941): 104.

14. See Clarke, *Professional Cinematography*, 127.

15. Hal Wallis, memo to Robert Lord, March 22, 1938, in the *Amazing Dr. Clitterhouse* files in the Warner Bros. Archives at USC.

16. Lang is quoted in Walter Blanchard, "Aces of the Camera XXIII: Charles Lang," *AC* 23 (Dec. 1942): 532.

17. Victor Milner, "Painting with Light," in *Cinematographic Annual*, vol. 1, ed. Hal Hall (Los Angeles: American Society of Cinematographers, 1930), 96.

18. Theodore Fuchs, *Stage Lighting* (Boston: Little, Brown, and Co., 1929), 10.

19. Milner, "Painting with Light," 96.

20. Charles Lang, "The Purpose and Practice of Diffusion," *AC* 14 (July 1933): 193.

21. John Arnold, "Why Is a Cinematographer?," *AC* 17 (Nov. 1936): 462.

22. Phil Tannura, "The Experimental 'B's,'" *AC* 22 (Sept. 1941): 420, 443. See also Herbert Aller, "Koffman's Mystery Effect Stills," *International Photographer* 10 (Sept. 1938): 13–16.

23. Milner, "Painting with Light," 96.

24. Walter Bluemel, "Composition in Practice, Part II," *International Photographer* 6.8 (Sept. 1934): 19.

25. Clarke, *Professional Cinematography*, 124.

26. Clarke, *Professional Cinematography*, 81.

27. Charles Lang, quoted in the microform *Charles Lang: An American Film Insti-*

tute Seminar on His Work (Glen Rock, N.J.: Microfilming Corporation of America, 1977), 11–12. In this quotation, the word "atmosphere" is prompted by a question from the crowd.

28. Joseph Ruttenberg mentions this detail in an interview, in Scott Eyman, ed., *Five American Cinematographers* (Metuchen, N.J.: Scarecrow Press, 1987), 42. Ruttenberg is referring to *A Day at the Races*, which was released in 1937, when the multiple-camera system was no longer the norm.

29. Victor Milner, "Creating Moods with Light," *AC* 15 (Jan. 1935): 14.

30. James Wong Howe, "Lighting," in *Cinematographic Annual*, vol. 2, ed. Hal Hall (Los Angeles: American Society of Cinematographers, 1931), 56.

31. Daniel B. Clark, "Composition in Motion Pictures," in *Cinematographic Annual*, vol. 1, ed. Hal Hall (Los Angeles: American Society of Cinematographers, 1930), 82.

32. Walter Bluemel, "Composition in Practice: Part II," *International Photographer* 6.8 (Sept. 1934): 18.

33. Wong Howe, "Lighting," 56.

34. John Arnold, "Background vs. Foreground Lighting," *AC* 17 (March 1937): 118.

35. Quoted in Leonard Maltin, *The Art of the Cinematographer: A Survey and Interviews with Five Masters* (New York: Dover, 1978), 108.

36. Wong Howe, "Lighting," 57.

37. Given the fact that cinematographers were fond of comparing their work to painting, it is worth mentioning a possible precursor to this approach: Leonardo da Vinci. According to John Shearman, Leonardo developed a strategy of lighting for aesthetic reasons like unity and modeling, while situating the scene in a location where those same lighting decisions would appear plausible; see John Shearman, "Leonardo's Colour and Chiaroscuro," *Zeitschrift für Kunstgeschichte* 25 (1962): 27.

38. Milner, "Painting," 91.

39. Milner, "Painting," 91.

40. An article by Lewis Physioc also discusses the spatial benefits of the cast shadow. They add pictorial interest, but they also reveal the forms of certain objects. Physioc illustrates this point by showing the same shadow cast upon triangular, rectangular, and cylindrical columns. The cast shadow clearly reveals the shape of each column. See Lewis Physioc, "More About Lighting," *International Photographer* 8 (Aug. 1936): 4–5. For an excellent art historical account of these issues, drawing useful distinctions among different kinds of shadows, see Michael Baxandall, *Shadows and Enlightenment* (New Haven: Yale University Press, 1995), 1–15.

41. See David Bordwell, "Glamor, Glimmer, and Uniqueness in Hollywood Portraiture," *Hollywood Glamour, 1924–1956: Selected Portraits from the Wisconsin Center for Film and Theater Research* (Madison: Elvehjem Museum exhibit catalogue, 1987).

42. Clarke, *Professional Cinematography*, 158.

43. George Folsey, "Some Don't, But, I Like Light Sets," *AC* 14 (May 1933): 15, 36.

44. Harry Burdick, "Valentine's Technique Is Vivid and Modern," *AC* 17 (Sept. 1936): 372, 379.

7. The Art of Balance

1. John Arnold, "Why Is a Cinematographer?," *AC* 17 (Nov. 1936): 462.
2. Quoted in Charles Higham, *Hollywood Cameramen: Sources of Light* (Bloomington: Indiana University Press, 1970), 75.
3. Charles B. Lang Jr., "The Purpose and Practice of Diffusion," *AC* 14 (Sept. 1933): 193–194.
4. Quoted in Leonard Maltin, *The Art of the Cinematographer: A Survey and Interviews with Five Masters,* enl. ed. (New York: Dover, 1978), 108.
5. Charles G. Clarke, *Charles Clarke's Professional Cinematography* (1964; Hollywood: American Society of Cinematographers, 2002), 115.
6. William Stull, "Concerning Cinematography: Critical Comments on Current Pictures," *AC* 13 (June 1932): 23.
7. Stull, "Concerning Cinematography," 23.
8. Thalberg is quoted by Karl Struss in the microform *Karl Struss: An American Film Institute Seminar on His Life and Work* (Glen Rock, N.J.: Microfilming Corporation of America, 1977), 32–33. See also Scott Eyman, ed., *Five American Cinematographers* (Metuchen, N.J.: Scarecrow Press, 1987), 4. In the Eyman interview, Struss claims that Thalberg told him this during the filming of *Ben-Hur* (1926).
9. This story is told on page 3 of an unsigned essay entitled "James Wong Howe," in the ASC file on Howe at the Margaret Herrick Library in Los Angeles. The film in question is *Whipsaw* (1936). The same library also holds the James Wong Howe collection. In folder 208 from this latter collection, there is an untitled and undated article by Howe himself. In this article Howe writes, "One big studio attracts cameramen who like to make everything look frothy and beautiful. If they don't, the studio interferes to bring that state about. I know, I was fired for refusing when I considered the story action demanded realistic photography" (4–5). Howe does not name the studio, but he is probably referring to his experiences at MGM, while working on *Whipsaw*. Soon after that film Howe left MGM. A few years later he began to work for Warner Bros.
10. Quoted in Maltin, *Art of the Cinematographer*, 101–102.
11. Clarke, *Professional Cinematography*, 51.
12. Alain Silver, "An Oral History of James Wong Howe" (1969), 63–64. This document can be found in folder 213 of the James Wong Howe collection at the Margaret Herrick Library.
13. Victor Milner, "Painting with Light," in *Cinematographic Annual*, vol. 1, ed. Hal Hall (Los Angeles: American Society of Cinematographers, 1930), 95.
14. Victor Milner, "Creating Moods with Light," *AC* 15 (Jan. 1935): 15.
15. Milner, "Creating Moods," 15.

16. Milner, "Creating Moods," 14.

17. Laura Mulvey, "Visual Pleasure and Narrative Cinema," in *Narrative, Apparatus, Ideology*, ed. Philip Rosen (New York: Columbia University Press, 1986), 198–209.

18. This discussion of Hawks draws on the famous chapter on the auteur theory in Peter Wollen, *Signs and Meaning in the Cinema* (Bloomington: Indiana University Press, 1972), 74–115.

19. Joseph Walker and Juanita Walker, *The Light on Her Face* (Hollywood: American Society of Cinematographers, 1984), 229.

20. Quoted in Higham, *Hollywood Cameramen*, 40–42.

21. Historians often cite this passage as evidence that cinematographers were influenced by painterly ideals; see, for instance, Peter Baxter, "On the History and Ideology of Film Lighting," *Screen* 16.3 (autumn 1975): 101–104; and Richard Dyer, *White* (New York: Routledge, 1997), 118. However, it should be noted that Garmes is using the term in an idiosyncratic way, referring to a light from above. Though painters like Rembrandt did light from above, a more accurate definition of the term "north-light" would be "soft light," since a north-facing window in the Northern Hemisphere receives indirect sunlight—like the soft light of bounced lighting. On this point I would agree with Barry Salt, who uses "soft light" as a synonym for "north-light"; see Barry Salt, *Film Style and Technology: History and Analysis*, 2nd ed. (London: Starword, 1992), 326.

22. Quoted in Higham, *Hollywood Cameramen*, 57, 70.

23. Fabrice Revault d'Allonnes, *La lumière au cinéma* (Paris: Cahiers du cinéma, 1991).

24. Ernst Gombrich, "Norm and Form: The Stylistic Categories of Art History and Their Origins in Renaissance Ideals," *Gombrich on the Renaissance*, vol. 1: *Norm and Form* (London: Phaidon, 1966), 81–98.

25. David Bordwell, Janet Staiger, and Kristin Thompson, *The Classical Hollywood Cinema: Film Style and Mode of Production to 1960* (New York: Columbia University Press, 1985). See esp. Bordwell's chapter, "The Bounds of Difference," 70–84.

26. I make a more extended argument along these lines in Patrick Keating, "Emotional Curves and Linear Narratives," *Velvet Light Trap* 58 (fall 2006): 4–15.

27. Lee Garmes, "Photography," *Behind the Screen: How Films Are Made*, ed. Stephen Watts (London: A. Barker, 1938), 107–108.

28. See Higham, *Hollywood Cameramen*, 37–38.

8. The Promises and Problems of Technicolor

1. Quoted in Leonard Maltin, *The Art of the Cinematographer: A Survey and Interviews with Five Masters*, enl. ed. (New York: Dover, 1978), 68.

2. Edward Buscombe, "Sound and Color," in *Movies and Methods*, vol. 2, ed. Bill Nichols (Berkeley: University of California Press, 1985), 83–92. Steve Neale

develops and complicate Buscombe's ideas in *Cinema and Technology: Image, Sound, Color* (London: Macmillan, 1985).

3. Scott Higgins, *Harnessing the Technicolor Rainbow: Color Design in the 1930s* (Austin: University of Texas Press, 2007).

4. L. T. Troland, "Some Psychological Aspects of Natural Color Motion Pictures," *TSMPE* 11.32 (Sept. 1927): 684.

5. Troland, "Psychological Aspects," 685–686.

6. Natalie M. Kalmus, "Color Consciousness," *JSMPE* 25.2 (Aug. 1935): 139–140.

7. Kalmus, "Color Consciousness," 142.

8. Kalmus, "Color Consciousness," 147.

9. Higgins, *Technicolor Rainbow*, 19–20.

10. David O. Selznick to Henry Ginsberg, March 27, 1936, Selznick Files, *The Garden of Allah*–Cameramen. David O. Selznick Collection, Harry Ransom Humanities Research Center, University of Texas at Austin (this collection of documents will be referred to as the Selznick Files).

11. David O. Selznick to Henry Ginsberg, April 14, 1936, Selznick Files, *The Garden of Allah*–Cameramen.

12. Boleslawski's response is handwritten on the Selznick–Ginsberg memo of April 14, 1936. Incidentally, Rosson had already fallen in love with one Hollywood star: he was married to Jean Harlow from 1933 to 1935.

13. David O. Selznick to Henry Ginsberg, undated, Selznick Files, *The Garden of Allah*–Cameramen. The content of the memo suggests that it was written after April 14, 1936.

14. David O. Selznick to Ray Klune and Victor Fleming, March 13, 1939, Selznick Files, *Gone with the Wind*–Technicolor. At the end of the message, Selznick indicates that he wants all of his departments to discuss this memo.

15. See David O. Selznick to Henry Ginsberg and Ted Butcher, Oct. 18, 1938, Selznick Files, *Gone with the Wind*–Cameramen.

16. David O. Selznick to Henry Ginsberg, May 3, 1939, Selznick Files, *Gone with the Wind*–Cameramen.

17. C. W. Handley, "Lighting for Technicolor Motion Pictures," *JSMPE* 25.6 (Dec. 1935): 429. See also Handley, "The Advanced Technic of Technicolor Lighting," *JSMPE* 29.2 (Aug. 1937): 173.

18. Quoted in Maltin, *Art of the Cinematographer*, 69.

19. J. A. Ball, "The Technicolor Process of Three-Color Cinematography," *JSMPE* 25.2 (Aug. 1935): 134.

20. Richard Dyer, *White* (New York: Routledge, 1997), 90.

21. Dyer, *White*, 127.

22. Ball, "Technicolor Process," 133.

23. Troland, "Psychological Aspects," 687.

24. Winton Hoch, "Technicolor Cinematography," *JSMPE* 39 (Aug. 1942): 98.

25. Rouben Mamoulian, "Some Problems in Directing Color Pictures," *JSMPE* 25.2 (Aug. 1935): 151.

26. Herbert Kalmus to Henry Ginsberg, April 24, 1936, Selznick Files, *The Garden of Allah*–Technicolor.

27. Quoted in Charles Higham, *Hollywood Cameramen: Sources of Light* (Bloomington: Indiana University Press, 1970), 33–34.

28. Quoted in Higham, *Hollywood Cameramen*, 34.

9. The Flow of the River

1. André Bazin, "The Evolution of the Language of Cinema," in *What Is Cinema?*, vol. 1, ed. and trans. Hugh Gray (Berkeley: University of California Press, 1967), 31.

2. Michael Baxandall, *Patterns of Intention: On the Historical Explanation of Pictures* (New Haven: Yale University Press, 1985), 58–62.

3. Edward Weston, "Random Notes on Photography" (1922), in *Photography: Essays and Images*, ed. Beaumont Newhall (New York: Museum of Modern Art, 1980), 223.

4. Edward Weston, "Photography—Not Pictorial" (1934), in *Photographers on Photography*, ed. Nathan Lyons (Englewood Cliffs, N.J.: Prentice-Hall, 1966), 155.

5. Quoted in John Paul Edwards, "Group f.64" (1935), in *Photography: Essays and Images*, ed. Newhall, 253. See also Alan Trachtenberg, *Reading American Photographs: Images as History, Matthew Brady to Walker Evans* (New York: Hill and Wang, 1989), 256.

6. Paul Strand, "Photography and the New God" (1922), in *Photographers on Photography*, ed. Lyons, 139.

7. See Joel Eisinger's discussion of Strand's theory of photography in *Trace and Transformation: American Criticism of Photography in the Modernist Period* (Albuquerque: University of New Mexico Press, 1995), 56–59.

8. Trachtenberg, *Reading American Photographs*, 190–191.

9. See Eisinger, *Trace and Transformation*, 87–90.

10. Edward Steichen, "The FSA Photographers" (1938), in *Photography: Essays and Images*, ed. Newhall, 267.

11. Anonymous, "Speaking of Pictures," *Life* 3.26 (Dec. 27, 1939): 4–6.

12. Anonymous, "Speaking of Pictures," *Life* 2.6 (Feb. 8, 1937): 4–5.

13. Anonymous, "Berenice Abbott Photographs the Face of a Changing City," *Life* 4.1 (Jan. 3, 1938): 40–45.

14. Anonymous, "The South of Erskine Caldwell Is Photographed by Margaret Bourke-White," *Life* 3.21 (Nov. 22, 1937): 48–52.

15. Anonymous, "Life on the American Newsfront," *Life* 2.7 (Feb. 15, 1937): 14–17.

16. See the advertisement in *Life* 2.20 (May 17, 1937): 58–59.

17. Anonymous, "How Starlets Learn to Act," *Life* 3.20 (Nov. 15, 1937): 36–41.

18. Anonymous, "Speaking of Pictures," *Life* 2.1 (Jan. 4, 1937): 6.

19. Karl Struss, "Photographic Modernism and the Cinematographer," *AC* 15 (Nov. 1934): 296.

20. Struss, "Photographic Modernism," 296.

21. William Stull, "Camera Work Fails True Mission When It Sinks Realism for Beauty," *AC* 18 (February 1938): 56, 59–60.

22. See William Paul, *Ernst Lubitsch's American Comedy* (New York: Columbia University Press, 1983).

23. Victor Milner, "Victor Milner Makes Reply to Ernst Lubitsch as to Realism," *AC* 18 (March 1938): 94–95.

24. See Charles Lang, "The Purposes and Practices of Diffusion," *AC* 14 (July 1933): 193; and John Arnold, "Why Is a Cinematographer?," *AC* 17 (Nov. 1936): 462.

25. Anonymous, "Lighting the New Fast Films," *AC* 19 (Feb. 1939): 70.

26. Thanks to James Naremore for pointing out this detail.

27. Gregg Toland, "Realism for 'Citizen Kane,'" *AC* 21 (Feb. 1941): 54.

28. Toland, "Realism," 55.

29. Bazin, "Evolution of the Language of the Cinema," 35–36.

30. Herb A. Lightman, "Mood in the Motion Picture," *AC* 28 (Feb. 1947): 49, 69. In the original article some of these lines are printed in the wrong order. I have corrected this to make the passage readable.

31. Robert L. Carringer, *The Making of Citizen Kane*, rev. and updated ed. (Berkeley: University of California Press, 1996), 83.

32. James Naremore, *The Magic World of Orson Welles*, rev. ed. (Dallas: Southern Methodist University Press, 1989), 78.

33. David Bordwell, Janet Staiger, and Kristin Thompson, *The Classical Hollywood Cinema: Film Style and Mode of Production to 1960* (New York: Columbia University Press, 1985), 243–261, 341–352.

34. Charles G. Clarke, "How Desirable Is Extreme Focal Depth?," *AC* 22 (January 1942): 36.

35. Clarke later became a cinematography teacher at UCLA. In his 1964 textbook he consistently argues that the cinematographer should think of the camera as representing the eye of the spectator; see Charles G. Clarke, *Charles Clarke's Professional Cinematography* (1964; Hollywood: American Society of Cinematographers, 2002), 87.

36. Years later Joseph Ruttenberg would make the same criticism; see the microform *Joseph Ruttenberg: An American Film Institute Seminar on His Work* (Glen Rock, N.J.: Microfilming Corporation of America, 1977), 71.

37. Clarke, "How Desirable?," 14.

38. For an account of *Bataan*'s representation of the group, see Jeanine Basinger, *The World War II Combat Film: Anatomy of a Genre* (Middletown: Wesleyan University Press, 2003), 34–57.

39. Herb A. Lightman, "'13 Rue Madeleine: Documentary Style in the Photoplay," *AC* 28 (March 1947): 89.

40. Quoted in Walter Blanchard, "Aces of the Camera VII: James Wong Howe," *AC* 22 (July 1941): 346.

41. James Wong Howe, "The Documentary Technique in Hollywood," *AC* 25 (Jan. 1944): 10, 32.

42. See Charles Higham, *Hollywood Cameramen: Sources of Light* (Bloomington: Indiana University Press, 1970), 78–79.

10. Film Noir and the Limits of Classicism

1. Janey Place and Lowell Peterson, "Some Visual Motifs of Film Noir," in *Film Noir Reader*, ed. Alain Silver and James Ursini (New York: Limelight, 1996), 64–75.

2. Paul Schrader, "Notes on Film Noir," in *Film Noir Reader*, ed. Silver and Ursini, 53–63.

3. Barry Salt, "From Caligari to Who?," *Sight and Sound* 48 (spring 1979): 119–123.

4. Thomas Elsaessar, *Weimar Cinema and After: Germany's Historical Imaginary* (New York: Routledge, 2000), 420–444.

5. Marc Vernet, "Film Noir on the Edge of Doom," in *Shades of Noir*, ed. Joan Copjec (New York: Verso, 1993), 1–31.

6. I borrow the term "mutually limiting demands" from Ernst Gombrich, "Raphael's Madonna della Sedia," *Gombrich on the Renaissance*, vol. 1: *Norm and Form* (London: Phaidon, 1993), 74. This essay is a nice companion piece to "Norm and Form," cited in ch. 7, n. 24.

7. Phil Tannura, "The Experimental 'B's,'" *AC* 22 (Sept. 1941): 420.

8. James Naremore, *More Than Night: Film Noir in Its Contexts* (Berkeley: University of California Press, 1998), 136–166.

9. Edward Dimendberg, *Film Noir and the Spaces of Modernity* (Cambridge, Mass.: Harvard University Press, 2004), ch. 1.

10. For two classic accounts, see Sylvia Harvey, "Woman's Place: The Absent Family of Film Noir," in *Women in Film Noir*, ed. E. Ann Kaplan (rev. ed.; London: British Film Institute, 1980), 22–34; and Frank Krutnik, *In a Lonely Street: Film Noir, Genre, Masculinity* (New York: Routledge, 1991).

11. Most of this biographical information is drawn from Todd McCarthy's introduction to John Alton, *Painting with Light* (1949; Berkeley: University of California Press, 1995).

12. Naremore, *More Than Night*, 174.

13. Alton, *Painting with Light*, 32. The italics are Alton's.

Index